IS INDIA BEING
COLONISED AGAIN?

IS INDIA BEING COLONISED AGAIN?

From Ship to Chip

Dr. Nishant Das

January 14, 2026

Copyright © 2026 Nishant Das

All rights reserved. No part of this book may be reproduced, stored, or transmitted in any form without prior written permission of the publisher.

ISBN: 979-8-218-91128-7

Published by Ambeone Press

First edition, 2026

This book is dedicated to the brave, patriotic and curious, which includes Pasha, Pari and their Servants.

CONTENTS

Preface i
Introduction ix

Parallels to the Past

1   Ship to Chip 3
2   Company Raj to Platform Raj 15
3   Trade Dependency to Technology Dependency 27
4   Colonial Monopoly to Digital Monopoly 43
5   Population Governance to Behavioural Analytics . . . 69
6   Newspaper to Social Media 93
7   Humiliation to Subordination 111
8   Cultural Programming to Algorithmic Socialisation . . 125
9   Colonial Subjects to Satisfied Consumers 145

Reckoning with the Present

10 The Reckoning 175
11 The Hierarchy of Innovation 187
12 Hope, Heroes, and Exceptions 193

Looking to the Future

13 Iron: The Gain & Loss of the Innovation Spirit . . . 209
14 Towards Digital Swaraj 217

References and Further Reading 221

PREFACE

As a child of the 1990s, I encountered India's colonial past and its post-independence struggles as distinct, faraway tales from a long-forgotten era. They felt archived, concluded, and seemingly unrelated to the present. Time insulated me from the horrors of that period.

The technological surge of the 1990s and the opening of markets reinforced that sense of distance. There was always a cool technology to buy, a new movie to watch, a recently opened mall to visit, and a newly announced video game to look forward to. Everything pointed forward. The future felt expansive. The future was exciting.

History thrilled, captivated, and engaged. But its value was mostly that of a well-told story, something to be heard at bedtime, seen in films, and, at times, quizzed over in history class.

In that sense, India's history felt more like myth, legend, and folklore. And like all stories, it carried within it small plot holes. One in particular always itched at the back of my mind, trivial yet persistent: why did the British leave?

The explanations offered all sounded plausible. None, however, seemed to fully account for why anyone would voluntarily relinquish something so vast, so valuable, and so meticulously constructed. Why would Britain willingly part with the jewel of its empire without a fight?

The question remained an unanswered curiosity. India's future, intertwined with my own sense of the future, had to be exciting.

Never did it occur to me that the future I was so looking forward to might in fact be laden with the bleakness and horrors of the past. India's past felt distinct, paid for, and settled. The future and the past felt completely disconnected. The idea that the past might return, that it might reassert itself in any form, never crossed my mind. Why would such a thought even come to mind?

And yet, as I moved through academia and into my professional life, my experiences and my growing understanding of the technologically enabled world we now inhabit began, slowly and in fragments, to answer that unanswered question. I began to understand why the British left.

The answer lay in a deeper understanding of technology. A key realisation is that technology is fundamentally the manipulation of fire and an amplifier of intent. Said to be gifted by the gods, humanity quickly learned to wield fire for its own uses and ambitions. It scared animals, forged blades, and built armies. As our mastery of fire grew subtler—evolving from the hearth to the piston to the microprocessor—technology advanced, and human society evolved with it.

It is a historical constant that technology grants its wielder power over others. Fire dazzles, warms, and protects. Because of its capacity to captivate and secure, others inevitably seek access to it. Within that seeking lies the seed of dependency. Those who control the "fire" can dictate terms, grant or deny access, and exercise leverage at will.

However, wielding power through technology is a rare art. It demands more than just possession; it requires vision, persistence, and a proactive relationship with the future, all of which is flavored by the intent of its wielder. While one wielder might use the "fire" of technology to illuminate a path toward national sovereignty, another may use it to cast a shadow of dependency.

This progression of power through technology largely shapes India's colonisation. The British learned to channel heat into steam, and steam into ships. With those ships, distance collapsed and trade accelerated. India was dazzled by this machinery, suddenly finding itself trading at an unprecedented scale. British steam was dependable, predictable, and quietly seductive.

But soon, fascination turned to dependence, and dependence to extraction. The British used their technological edge to rewire India's economic and social order—dictating what was grown, who traded, and who profited. They followed a precise empire-building sequence: exclusion, extraction, and enforced obedience. In the end, technological domination turned the trader into the colonizer.

India's independence arrived just as the world was entering a period of extraordinary scientific and technological advance. The strategic significance of these shifts would have been immediately apparent to the British, who had already demonstrated a capacity to convert technological advantage into global dominance.

What if such a technological leap enabled control without flags, administrators, or force? As technology becomes more refined, from steam to electrons, would those with the vision and intent to wield power not refine their methods as well? If this is the case, in such an evolving world, could power simply reposition itself, continuing the inner work of colonisation through technology, without the friction of symbols or the spectacle of overt rule?

Seen this way, the British departure from India may not signal the failure of empire, but the evolution of its method. One that no longer required physical presence.

As science and technology surged ahead in the West, India's inherited dependencies would be further solidified through its technological lag. Technology would deepen dependency and continue the work of extraction, ensuring that value created in India would still flow outward, even in the absence of visible rule.

In this configuration, India could become a nation of free citizens as long as it operated inside a world technologically engineered to treat them as subjects. The colonisers could leave, but colonial processes would remain.

Once I began to see the world in terms of processes rather than events, and systems rather than symbols, the British departure no longer appeared as the end of control, but as a change in its operating logic. The rulers had departed, but the logic of control had not. Technology would come to perform the work of subordination, quietly and invisibly.

What surprised me was how little of this had been explored. Colonialism had been studied as an event, a period, a political episode. It had rarely been examined as a process. Even more strikingly, the psychological processes underlying colonisation had remained largely unexplored. History concerned itself with dates, policies, administrators, and battles. It recorded what was done, but not how it was done, nor what it did to the inner landscape of the mind.

Colonialism revealed itself not as something that ended, but as a process, an operating logic, a sequence capable of evolution. I placed technology at the centre of India's historical colonisation process, tracing its role in enabling trade, economics, and extraction. I followed that logic into institutions, incentives, and organisational design. And finally, I descended into the least explored territory of all: the psychological and cultural processes through which subjugation was internalised, imagination imported, and aspiration constrained.

Through this exercise, I began to see uncanny parallels with processes at work today, shaping India and its psychological landscape. Not occasional exceptions or one-offs. The similarities with the past were structural, recurring, and deeply unsettling. That it was occurring without form, authority, or overt coercion made the consequences more chilling.

What I had always assumed to be safely confined to the past appeared to be re-emerging in the present. The comfort of historical distance began to erode. The past was no longer merely something that had happened. It was something that could happen again.

It was in this moment that I was compelled to write this book. Not as alarm, but as inquiry and examination. I place the argument before the reader as a series of questions, not a conclusion.

- Is India being colonised again?
- If so, is it occurring through process rather than through symbols?
- If it has, through what mechanisms does it now operate?
- Is India failing to recognise these mechanisms because they no longer resemble colonisation as it was once known?
- If India is not being colonised again, is the psychological imprint of the previous colonial logic still shaping behaviour, aspiration, and self-perception?
- Are we exercising genuine agency, or merely operating within constraints we have internalised?

One might object that if colonial control never fully disappeared but merely became more refined, then to speak of its "return" is conceptually flawed. Yet this objection overlooks the rupture introduced by technology itself. Each major technological shift resets the field. Old systems destabilise. Existing hierarchies are disrupted. New architectures of power become possible. In such moments, power cannot persist unchanged; it is forced to reconfigure itself.

This pattern is visible across every technological transition. VHS vanished in the age of streaming. Entire corporations collapsed when they failed to adapt.

In such moments of technological evolution, every nation must scramble, reposition, rebuild, and stake its claim within the new order. In this period of flux, there remains space for a country like India to seize power, become truly free, and shape its own future. This is especially true in the current age of AI, and will be even more so in the coming era of quantum and other transformative technologies. If such ruptures temporarily reset the field, the question becomes whether India will recognise the opportunity and reorient itself. It is in this sense that the question is posed: is India being colonised again, or is it failing to seize the moment?

INTRODUCTION

Colonialism is widely treated as a closed chapter. It is remembered as a cultural injustice, a political struggle, and a civilisational trauma. Heroes are celebrated, freedom is commemorated, independence is ritualised. The story is declared complete and disconnected from the future.

Why, then, reopen a chapter that has already been closed? Why revisit a story that is considered complete? Why ask if India is being colonised again?

The assumption that colonisation belongs to the past is comforting. It allows the injustice, indignity, and humiliation of that era to be archived, moralised, and safely forgotten.

And indeed, if colonisation were truly in the past, it would be best left there. But what, exactly, must be missing in the present for us to say that colonisation belongs to the past? Which specific markers of colonialism do we look for, the absence of which allows us to declare that it no longer exists?

It would be naïve to expect colonialism to return in the same costume as last time. There will be no ships, no cannons, no forts, no governors, no rebellion, no flag. Those were the visible symbols of an earlier system. To wait for them is to misunderstand how power evolves.

We must look for process, not pageantry.

To view colonialism as a process is to practise real vigilance. It is to understand how power emerges, how it operates, how it expands, and how it subjugates. That is where the real story lies. And that story does not have an end. The story of power is not episodic. It is continuous. It is forever. It is eternal.

India, it seems, has not examined this story as extensively as it should. Its history is rich in colonialism as event and symbol, but thin on colonialism as process and system. If it had, it would not assume the chapter closed.

For processes of colonisation are returning—reorganised, disguised, relabelled, but intact in purpose. And India appears largely unprepared for it.

Colonialism and power are inseparable from technology, or more precisely, from innovation asymmetry. Technology is the material expression of intent: the organised application of knowledge, tools, and systems to reshape reality at scale. From time immemorial, from fire and the wheel to the steamship and the silicon chip, technology has extended human capacity. It has enabled survival, amplified strength, increased efficiency, expanded scale, stabilised coordination, and transformed environments.

Technology reduces uncertainty, and humans are wired to consistently gravitate toward what makes life predictable and certain. This concentrates power in the hands of those who control access, allowing them to set terms, regulate entry, and exercise leverage.

Innovation is the exploration of the unknown to create new technological capability. In other words, it is the capacity to acquire superior tools, methods, and systems. As innovation asymmetry grows, one society's ability to advance technologically inevitably becomes another's point of dependency. Where such asymmetry persists, leverage emerges. And where leverage grows, colonisation follows.

The process of power through technology follows a predictable pattern. It follows a sequence: innovation creates new capability; diffusion spreads it; as absorption increases, dependence forms around it; lock-in stabilises it; leverage is then exercised; and, over time, control is internalised psychologically.

There is no moralisation in this process. It is systems logic. It is how every durable power structure in history has operated, regardless of era, culture, or intent.

Moral judgement and ethical intent arise only in how power is exercised. They depend on whether the wielder seeks to expand human capability or to concentrate advantage through extraction and self-enrichment.

Consider a vaccine developed in the midst of a widespread outbreak. The country that creates it immediately acquires leverage. Others will seek access, urgently and at almost any cost. This is not exceptional. It is structural. Innovation creates technological capability. Capability creates dependency. Dependency creates power. This is the process this book examines.

How that power is exercised—through profit maximisation, strategic bargaining, or open sharing—is a separate question.

Whether through steam, rail, finance, silicon chips, platforms, or algorithms, the sequence remains unchanged. Power emerges as a process. This process is the mechanical spine of dominance. This book examines India's relationship with power through this lens to ensure that the country's trajectory is not left to the mercy, preferences, or ethics of external actors.

Here, it is important to note that while the sequence of power remains constant, the way it manifests is shaped by context. It adapts to the technology of the moment, the speed of innovation diffusion, baseline conditions, and the needs it addresses. Steam moved slowly. Data moves instantly. Vaccines spread under urgency. Platforms spread through convenience. The symbols change, the dynamics vary, but the sequence does not.

Power rarely arrives through conquest alone. It enters through trade, adoption, utility, and dependence. It accumulates gradually, across legal, economic, financial, and institutional layers. It does not announce itself. It normalises itself. And over time—sometimes decades, sometimes centuries—it consolidates. The process is adaptive and patient. The outcome is predictable.

When colonial history is viewed as process rather than episode, the parallels of the present become unmistakable. Trading companies have become platforms. Governors have become algorithms. Territory has become data. Labour has become attention. Ports has become cloud. Statute has become standard. The architecture has not changed; only its surface has.

Even the psychological processes of colonialism are now reappearing. These were not accidental by-products of rule, but integral mechanisms of control. Innovations in their own right. Britain did not merely extract resources from India; it extracted agency. It redirected aspiration, hollowed out confidence, and normalised subordination.

Possessing overwhelming technological and organisational superiority, Britain had the capacity to develop, to educate, to industrialise, and to elevate. Instead, it reorganised Indian society to serve an extractive function. This reorganisation was achieved through psychological conditioning that made extraction administratively efficient and socially stable. India was not merely governed. It was cognitively and socially reconfigured.

India's tragedy was not that Britain behaved this way. It was that India was structurally vulnerable enough to allow it.

The same psychological processes are now reappearing, in subtler and more pervasive forms. The West continues to shape Indian aspiration, define prestige, calibrate legitimacy, and train desire. External validation remains the default threshold of worth. Internal standards and definitions are still imported. India continues to view and judge itself through external eyes.

This is where power finally settles—not in territory or law, but in the architecture of the mind. Carefully engineered and coerced in the past, it is now achieved through algorithmic nudging, content curation, and preference shaping. The update in colonialism, from steam to chip, is that what was once imposed is now desired. And what was once resisted is now eagerly paid for. Rent extraction has become digital.

And as these processes continue to operate, India continues to import. Unlike other formerly colonised nations that confronted their "century of humiliation" with strategic resolve, India remains largely aloof. It adopts rapidly, scales enthusiastically, and executes efficiently—but rarely authors. It consumes architectures it did not design, operates standards it did not set, and builds upon foundations it does not control.

The talent is world-class. The market is massive. Adoption is unmatched. Yet agency remains thin. India boasts its market size as leverage, when in reality it is exposure. Scale without authorship is not power; it is vulnerability. Instead of converting talent into original capability, it continues to import. Instead of building, it buys. Instead of defining, it adopts.

India celebrates Western technology as progress, mistaking participation for advancement, while failing to recognise the precarity it is creating for itself. This is the Indian paradox: intense motion without strategic direction, and participation without power.

The question this book asks is not whether the West is exploiting India. That is a shallow and ultimately unproductive inquiry. The real question is whether India is being colonised again—or whether it is failing to seize the moment when power is being restructured.

This is not a story of external aggression. It is a story of internal unpreparedness. The AI era is reshaping the configuration of power, and the technological regimes that follow, from advanced automation to quantum systems, will further destabilise old hierarchies. New systems are being written. In such moments, nations either author the future or become footnotes in someone else's design. This book is concerned with whether India is choosing to write or is quietly choosing to import.

This book is structured as an inquiry, not a proclamation. Part I returns to history not to relive it, but to extract the operating logic of power. Each chapter isolates a step in the process of power—innovation, diffusion, dependence, lock-in, leverage, and internalisation—and traces how these mechanisms first appeared in India's colonial experience. The objective is not to mourn the past, but to recognise its patterns and confront their persistence in the present.

Part II consolidates what has been established in Part I and brings those mechanisms forward into the present. It examines the depth of India's technological capability, including what it innovates, what it celebrates, what it can build, and what it authors. On this basis, it revisits the central question of the book.

Here, the reader is invited to reckon with the present. To accept responsibility. And to decide, with deliberate resolve rather than passive comfort, how the future will be met.

Part III steps outside linear time in order to reorient perspective and clarify how the future must be addressed. It uses the story of iron to show how technological mastery shifted westward while India's own capability eroded, illuminating the conditions under which innovation is lost and consolidated. The story is both historical and symbolic. The book ends with a way forward.

Part I

PARALLELS TO THE PAST

1

From SHIP to CHIP

1.1 TECHNOLOGY MAKES EMPIRES

Empires are not sustained by territory alone but by control of the technology that defines an era. In every historical cycle, one form of innovation extends a nation's reach beyond its borders and determines the structure of global exchange. For the British Empire, it was the steam-powered ship. For today's digital powers, it is the silicon chip. The engine of domination has shifted from maritime propulsion to computational precision, but the logic remains the same: those who invent, own, and scale the dominant technology set the rules of trade, knowledge, and culture; those who do not must live by them.

The ship gave Britain mastery over the seas and control of the world's supply routes. The chip gives a handful of nations and corporations mastery over information flows, the new arteries of civilization. Where ships once carried goods, chips now carry logic; where steam propelled the imperial fleet, data propels the digital order and directs human attention. Power lies not in geography or manpower but in innovation itself, the capacity to translate knowledge into leverage. The shift from ship to chip is not merely technological but civilizational: power now lies in shaping how billions perceive, desire, and decide.

This shift marks the beginning of a new but familiar cycle of dependency. As before, technology does not merely expand markets; it defines who governs them. Nations that fail to create or control foundational technologies inevitably become consumers, not authors, of their own destiny. India, and much of the Global South, stands again at this historical inflection: will it command the tools of its digital future or repeat the pattern of dependency it once escaped?

1.2 INNOVATION ASYMMETRY: HOW EMPIRES ARE BORN

The Industrial Revolution turned the ship from a vessel of commerce into an instrument of empire. Steam engines allowed goods, troops, and ideas to move faster than ever before. This velocity translated directly into power. Innovation asymmetry, where one side invents and the other adapts, created a structural imbalance that reshaped the world. The West innovated; its colonies adapted. Britain's innovation advantage was not accidental. Its industrial ecosystem, coal, iron, machine tools, finance, and institutional rights, formed a complete loop from invention to deployment. Colonies like India provided the raw materials but were systematically excluded from the loop of creation. This division of roles, inventor and adapter, set the pattern for a century of dependency and extraction. As economic historians David Landes and Joel Mokyr have observed, innovation becomes imperial when the inventor's capacity to improve outpaces the adapter's capacity to replicate. Once that gap widens, the flow of value reverses, setting off a chain of dependency that transforms economic exchange into subordination.

Innovation asymmetry soon became economic asymmetry. The mastery of movement and production turned into mastery of trade and policy. Once technological control was secured, political intervention followed naturally, justified as "trade protection" but driven by the need to secure resources. The East India Company offers a telling example. To ensure a steady supply of tea for British consumers, it forcibly transformed India's agrarian landscape. After the early 1800s, tea cultivation, once native to China, was transplanted into Assam and Darjeeling through state-backed monopolies and indentured-labour systems. The colonised were compelled to adapt, not merely to new economic terms, but to an entirely foreign technological and agricultural order. The same steam engine that propelled British ships across oceans now powered an empire's appetite for control. The innovation that once symbolised progress now justified coercion. The spirit of invention, once creative, became extractive. The innovator became the coloniser; the adapter, the colonised.

Today's digital order mirrors that asymmetry almost exactly. The West, led by the United States, invents the algorithms, architectures, and platforms that shape the global information ecosystem. The rest of the world, including India, adopts and integrates them, adding scale but not authorship. Just as industrial Britain imported raw cotton and exported finished textiles, today's digital powers import raw data and export finished software. The pattern has changed only in medium, not in meaning. The economic consequences of this new asymmetry, from value capture to dependency through data, are only beginning to unfold, and I will explore them in later chapters.

1.3 TECHNOLOGY DEPENDENCY AS THE MECHANISM OF CONTROL

Scholars of technological sovereignty, from Shoshana Zuboff's Surveillance Capitalism to Manuel Castells' Network Society and Nick Srnicek's Platform Capitalism, have all traced how infrastructure becomes ideology. Their work frames technology not as neutral utility but as architecture of power. This book extends that lineage by arguing that digital dependency is not merely economic but civilizational: it determines whose cognition defines collective reality. India's struggle for Digital Sovereignty therefore belongs to the same continuum as earlier quests for political and economic independence, only fought now in code rather than colonies.

In both eras, control operates through infrastructure. Ships gave Britain control of maritime routes; chips give digital powers control of information routes. Infrastructural determinism, to borrow from McLuhan and Castells, suggests that whoever owns the channels of transmission ultimately controls the content and the terms of exchange. Trade routes once defined the geography of empire; now information routes define its topology. The steamship's advantage was not simply speed but reliability; it allowed predictable, repeatable exchange. Similarly, the semiconductor's advantage lies not merely in computation but in integration; it embeds control into every layer of digital life. From cloud networks to payment systems, every act of communication or commerce today flows through circuits and code that trace back to a few innovation hubs. The infrastructure is global, but its control is local, concentrated in corporate and national centres far from where most of the data originates.

In the nineteenth century, Britain's control over shipping lanes and telegraphs gave it leverage over colonies even without direct military presence. In the twenty-first, control over operating systems, app stores, and cloud APIs plays the same role. Access, not occupation, defines dominance. The modern empire doesn't extract taxes; it extracts rents from usage. Every digital interaction, a click, a query, a transaction, pays tribute through data.

Dependency does not emerge from scarcity of resources but from scarcity of technological autonomy. Countries rich in raw materials or human capital can still find themselves subordinated if they lack command over the means of computation. This is the essence of technonationalism, the idea that national power depends on ownership of key technologies and control of their value chains. In the industrial age, that meant coal and steel; in the digital age, it means chips, data centres, and algorithms. Each stage of the technological process carries unequal leverage. Design and standards capture the highest value; assembly captures the least. When a nation imports finished technology but exports raw inputs, whether minerals, code, or data, it remains at the bottom of the hierarchy even if its exports grow in volume. The dependency is structural, not transactional.

This asymmetry is reinforced by network effects. Once global standards are set, late entrants must conform to them or be excluded. Switching costs make autonomy progressively harder. The more a nation integrates foreign cloud systems, payment rails, or software stacks, the tougher it becomes to regulate or innovate independently. Dependency becomes embedded in infrastructure, invisible, irreversible, and cumulative. True sovereignty lies not in regulation but in capability. Governance and regulation without technological command is performance, not power. Laws may define rights, but code defines today's technology-embedded reality, and technology always moves faster than regulation. To legislate without the means to compute is to write rules in sand, washed away by each new wave of innovation. The real question for any modern nation is not who governs, but who compiled the code, not who sets policy, but who designed the algorithm. In a world where algorithms shape decisions long before laws can catch up, sovereignty begins not in parliament, but in source code.

1.4 TECHNOLOGY ABSORPTION WITHOUT AUTHORSHIP

India sits at a paradoxical crossroads. It is both a technology hub and a technological colony, a producer of talent but not of tools. Its engineers build code for the world, but the architectures they work within are foreign. Its markets are vast, yet its platforms are largely imported. This is not due to lack of intelligence or ambition but to a structural comfort with adoption over invention.

Consider a simple mirror from history: in 1850, India exported cotton and imported Manchester cloth; in 2025, it exports raw data and imports Californian algorithms. The form has changed, the dependency pattern has not. Nearly 90 percent of India's semiconductor demand and most of its cloud infrastructure still originate abroad, an echo of the nineteenth-century trade imbalance disguised in digital form. These figures are less about economics than about authorship: who writes the logic that others live by. The 2025 Microsoft–Nayara incident was a small but symbolic reminder of this vulnerability. When a global corporation could unilaterally suspend a major Indian company's digital access, it revealed how deeply dependent the nation's infrastructure had become on external systems. The denial was later reversed, but the message remained: sovereignty that relies on another's code is conditional. This is not an exception; it is the default configuration of the Indian digital ecosystem.

India's transformation from a resource exporter under colonial rule to a data exporter under digital capitalism marks continuity, not rupture. Its raw data, user behaviour, transactions, communications, feeds foreign algorithms that learn from Indian patterns and monetise global insights. In return, India consumes the finished digital goods: recommendation engines, language models, fintech apps. The imbalance of creation and consumption persists; only the commodity has changed.

The government's digital public infrastructure initiatives, such as Aadhaar, UPI, and ONDC, show that state-led innovation is possible, but these successes are still anchored on global hardware, cloud, and software dependencies. The base layer remains foreign, even when the interface appears domestic. India's innovation sovereignty is partial: it builds on top of systems it does not own.

1.5 THE COLONIAL REFLEX

Technological dependency sustains itself not only through infrastructure but through mindset. Centuries of colonial hierarchy have shaped what might be called the colonial reflex, a psychological comfort with dependency, the belief that progress comes from across the sea. Prestige is attached to adoption, not authorship; mastery is equated with Western validation, not creation. The result is a society that celebrates use but hesitates to invent.

This mindset expresses itself culturally and economically. Imported devices confer status; local innovations are treated as compromises. The same reflex that once glorified British textiles now glorifies Western smartphones. The underlying psychology is not admiration but insecurity, the learned assumption that originality lies outside one's borders. This is innovation aversion, a conditioned avoidance of originality, where safety lies in imitation and prestige in adoption.

The educational system reinforces this loop. Students are trained for accuracy, not imagination; risk is penalised more than failure is analysed. A young engineer who wants to start a company faces more social scepticism than one who joins a multinational or dreams of leaving for a job abroad. The colonial ship that once carried raw materials also carried the children of India's elite to British universities, where status was conferred through proximity to power. Two centuries later, the pattern endures. The same voyage now takes place through code and careers. The ship has become digital, ferrying talent to Western firms and universities that continue to define what "global" excellence means. In behavioural terms, it is a form of collective learned helplessness, a condition where capacity exists but only finds its wind abroad, setting sail not toward creation, but toward recognition.

Breaking this cycle requires more than policy incentives; it requires cultural reorientation. A nation cannot innovate if its citizens seek safety in imitation, adaptation, or jugaad. Just as the Industrial Revolution was fuelled by an entrepreneurial ethos, a willingness to build, fail, and rebuild, a digital renaissance demands psychological sovereignty. Innovation begins not in labs but in mindsets.

1.6 TOWARDS DIGITAL SWARAJ : SAIL YOUR OWN SHIP

If the ship once defined control of land and trade, the chip now defines control of information and cognition. The empire of the past was visible, marked by borders, flags, and armies. The empire of today is invisible, encoded in protocols, platforms, and permissions. Both operate on the same logic: innovation creates dependency, and dependency invites governance.

Digital Sovereignty begins with recognizing this continuity. Technological independence requires authorship. A nation that merely consumes foreign innovations participates in its own subordination. True sovereignty lies in the ability to design, build, and question the systems that shape its reality. This requires aligning manufacturing, software, institutions, and mindsets, the full stack of capability.

The path ahead is neither protectionist nor utopian; it is pragmatic. Build what you can, collaborate where you must, and never surrender the ability to create. Even the smallest act of creation matters, especially when it is easier to import. Sovereignty in the digital era demands authorship across the entire technological spectrum, from the base languages that define logic to the chips that execute it. Every layer outsourced is a layer of control relinquished. The code we write, the tools we compile, and the hardware we fabricate are not merely instruments of productivity but expressions of autonomy. India must, therefore, revisit the full stack of its technological dependence, from infrastructure to instruction, and rebuild it, piece by piece, on its own terms. A nation that failed to sail its own ships must not fail to write its own code.

2

From COMPANY RAJ *to* PLATFORM RAJ

2.1 LEGITIMISING PROFIT MAXIMISATION

On the last day of 1600, Queen Elizabeth I signed a charter that would change the world. It granted a handful of London merchants not just the right to trade with the East, but the exclusive right and royal protection to do so. This was a legal monopoly backed by the full authority of the English Crown. For the first time in modern history, a single document created a private corporation endowed with the powers of a state. The East India Company was not a state enterprise. It was something far more radical: a profit-driven entity that could wage wars, levy taxes, dictate policy, and govern territorie s, all while its ultimate purpose remained commercial gain.

A single piece of parchment had transformed a trading firm into a sovereign instrument. And yet, it was never the sovereign itself. That was the genius of its design.

The Crown could appear righteous, distant, moral, and above the fray, while the Company carried out the ruthless work on the ground. The Queen's seal provided legitimacy and cover; the Company's ledgers delivered the exploitation. When famine followed misrule, or when local populations suffered under heavy taxation, London could point the finger at "commerce" rather than the Crown.

Empire had discovered a powerful innovation: privatizing its own guilt. The sovereign could rule without direct responsibility and profit without blame.

It was not physical innovations like the steam engine or the musket alone that built the empire. What truly enabled it was what followed them: the legal obfuscation that made exploitation appear innocent. The separation of state from company through the Royal Charter was moral engineering: a second-layer invention that purified intent while delegating brutality.

Like priests blessing conquest in the name of faith, the Crown provided the moral cover, absolving the sovereign of direct sin, while the Company carried out the ruthless work on the ground. Power could act without guilt. Violence could proceed without visible consequence. It was a masterpiece of cunning. An innovation for profit maximisation so ruthless that it legitimised domination and systematised exploitation. While the machinery of domination was led by physical innovation, its legitimacy was born of invisible innovation.

2.2 FROM ROYAL CHARTER TO SECTION 230

Four centuries later, the charter returns in digital form. No wax seal. No monarch. Just a checkbox that says "I agree."

The new equivalent of the Royal Charter is hidden in plain sight: the document of Terms and Conditions that governs our digital lives. Each one grants corporations the privileges once reserved for the companies and agents of empire, and more crucially, the legal abstraction that makes those privileges appear harmless: to extract, to surveil, and to rule in the name of providing a service.

It is a compact between the individual and a corporation so vast it functions like a jurisdiction of its own. Few read it. Fewer question it. Yet all must accept it. Each click is an act of quiet submission, granting private platforms the kind of privileges once conferred by the Crown to its chartered merchants: dominion over markets, subjects, and now, minds.

What began as a royal signature has evolved into a daily ritual of consent. The user believes they are merely accessing a service. In truth, they are renewing a modern charter, one that binds them to rules they did not write, data they do not own, and power they cannot revoke. Delegated Sovereignty and Moral Distance In law, this is called consent. In practice, it is delegation: the quiet transfer of sovereignty from citizen to system, bypassing local governments entirely.

What was once the monopoly of a Company is now the immunity of a platform, protected by Section 230, a law that allows corporations to act without being held accountable for the consequences of their actions. Operating largely from the West, these companies are answerable to no one for what their networks unleash abroad. The governments that license them are equally insulated: the platforms are "private," the consequences "unintended."

The result is a chain of unaccountable power identical in structure to the old imperial design. The East India Company drew its authority from the Crown but disclaimed its politics: the Crown could claim innocence while the Company waged war. Today's platforms draw their legitimacy from Western law yet deny their geopolitics: Washington or Brussels can call it innovation while the platforms reshape societies.

The Company fired its muskets for profit while London called it trade. The modern platform triggers social movements, market panics, even the fall of governments, while the capitals of the West call it freedom.

If Section 230 is the new Royal Charter, then the platforms it protects are the new East India Companies: private powers governing public life in the name of innovation. It turns interference into benevolence, extraction into inclusion. It revives the old fiction of "gifting Western progress to the East." The West was not responsible for the East India Company firing bullets for profit: that was commerce, after all. Just as it is not responsible for the tweets that topple governments: that is merely free speech. Empire has shifted from cannon to code. But the moral distance remains the same.

The illusion of neutrality in platforms parallels the Charter itself. Both framed state-granted and state-defended private power as public service. As scholars like Nick Srnicek and Shoshana Zuboff have shown, platforms no longer operate within economies; they are the infrastructure that economies depend on. History shows that such dependence can always be leveraged. Section 230 perfects this design by detaching consequence from control, allowing corporations to act politically while claiming they are merely technological. They decide which voices rise and which vanish. Which leaders are silenced and which movements are amplified.

Yet when unrest follows, it is the local citizens who bear the fallout: the violence, the instability, the economic shock. Meanwhile, Western capitals reap the profit or the geopolitical leverage. The moral architecture remains the same: the West innovates to profit; the East adapts to survive.

2.3 SERVICE AS SEDUCTION

When the East India Company and other Europeans entered India, they came as service providers: merchants, not monarchs. In 1613, the Company received permission to trade and establish factories in port cities such as Surat. It brought curiosities from cold, faraway lands: Venetian glass, mirrors, silverware, clocks, muskets, wine, even beer. These were novelties that soon became status symbols among Indian elites.

In return, it carried away Indian cotton textiles, silk, indigo, and spices, sometimes to London, sometimes to Japan or Southeast Asia. This was the dawn of a cosmopolitan exchange system, a trading platform if one wishes, where European goods travelled east and Indian craft flowed across the world.

For Indian merchants, it was exhilarating. Their textiles and dyes were suddenly in demand in London, Amsterdam, even Edo Japan: markets they could never have reached alone. To see one's goods sail away to eager buyers at the far edges of the world must have been a matter of pride. Such a network offered what no caravan or coastal broker could: punctual ships, predictable routes, and contracts that spanned oceans. It seemed a new age of prosperity, a partnership between skill and scale.

Yet the pride was misplaced. The ownership of that prosperity lay elsewhere. The ships were British. The routes European. The buyers foreign. Indian merchants filled the orders but did not know the buyers of their goods. Had they owned the ships, they could have sailed their own routes, to Japan, to England, to wherever ambition led. Instead, they waited for the next steam-powered vessel to dock at their shores. India supplied. But it did not control.

The Indian trader rejoiced in the order's size, never realising that the order itself was the leash. Predictable trade is not the problem. As we will explore in the next chapter, it is the ownership of that predictability that determines who prospers and who obeys. When reliability itself is outsourced, dependence becomes a weapon: coordination turned into control.

2.4 THE MIDDLEMAN EMPIRE

The East India Company was not a producer. It did not till the fields. It did not weave the cloth. It did not mine the goods it traded. It was a mover. It profited not from what it made, but from what it moved. Its genius lay in controlling flow: routes, ports, tariffs, insurance. In effect, it became the first great platform of the modern age, connecting markets while quietly controlling them.

The Company's power came not from production but from permission: tariffs, customs, and routes that turned geography into profit. It was a logistical monarchy: mastery through movement, matching the seller to the buyer and carrying their goods across oceans. In essence, it was the first platform of its kind.

2.5 ENTRAPMENT BY DESIGN

The prosperity India seemed to enjoy was one-sided. During the seventeenth century, India was a net importer of silver and gold. European traders had little that India wanted; nearly ninety percent of the Company's exports from England to Asia were bullion. Such a drain of European metal was not an accident of taste. It was the opening investment in a longer algorithm of control.

The provisions of the Charter allowed the Company to play a longer game: to bear losses now for monopolistic profits later. The bullion that poured into Surat would soon return to London, multiplied a hundredfold in the form of taxes, rents, and tribute. To win India's trust and secure its position, the Company outbid its Dutch and Portuguese rivals, paying well above market for silk, indigo, and cotton.

Gradually it cornered the market and monopolised trade with India. For a while, the arrangement seemed divine: Indian traders earned more; the Company secured steady supply. What looked like generosity was entrapment, a loss ledgered in London to buy permanence abroad.

So while Indian merchants were focused on profits, the Company gained power. The Company's excellence in service, its punctual ships, secure contracts, and above-market prices, was not virtue but design. Creating dependency was a strategy of empire: a system built to embed reliance quietly before domination became visible.

By mid-century, the illusion of partnership had reached its perfect balance: India felt prosperous just as its autonomy was being hollowed out. The looms still hummed, the ports still glittered, and the merchants still counted profit, unaware that they were already working inside someone else's plan. Like a frog being slowly boiled, mistaking the initial gentle heat for comfort, India mistook the warmth of trade for the glow of prosperity.

By the time the British monopolised India's trade, escape was impossible. The seduction of service was complete, the dependency irreversible. With monopoly secured, the age of extraction was ready to begin.

2.6 THE MODERN MIRROR

Four centuries later, the same logic of service as seduction still runs, written not in ledgers but in algorithms. The platform, like the Company, begins by serving and ends by owning. What the ship once did for trade, the chip now does for choice. Both promise efficiency, and both deliver dependence: a seduction powered by convenience. The age of service that once preceded colonial extraction is repeating itself in digital form today.

Today's foreign platforms also come to serve. They enter markets not as rulers but as helpers, the middlemen of modern life. Across search, communication, commerce, and now intelligence itself, they promise convenience, not control: instant results, effortless transactions, intuitive interfaces, and 24-hour assistance. The tone is benevolent, not imperial. Everything about them is designed to appear seamless, to create trust through ease. Reliability, not rule, is the currency of persuasion.

Like the Company's punctual ships, platforms pride themselves on precision. Orders arrive on time; dashboards never fail; the app interface anticipates what you want before you search. The smoother the experience, the deeper the dependency. What feels like empowerment is the quiet discipline of infrastructure. Convenience becomes habit; habit becomes captivity.

And like the old empire, this one pretends to empower. The merchant was once praised for exporting India's craft;

the influencer today celebrates "reaching the world." But both operate within invisible permission. The seller owns neither store nor customer; the creator owns neither platform nor audience. The citizen logs in believing they enter a marketplace; in truth, they enter a jurisdiction.

The illusion is global participation. Just as Indian traders once rejoiced at seeing their goods sail to faraway markets, modern sellers and creators believe they "reach the world." They do — but only through routes they do not own. The servers, search indexes, payment gateways, and customer data remain firmly in the hands of the platform. To sell or speak globally is now to do so on borrowed infrastructure.

At its core, the logic has not changed. What the East India Company achieved with ships and charters, today's tech giants achieve with servers and subscriptions. Power lies not in production, but in positioning — in owning the movement between producer and consumer. Canvas then, code now. The medium has changed; the principle endures. Route ownership defines power.

Today, sellers, drivers, and content creators on foreign platforms call themselves "partners," just as Indian traders once mistook themselves for equals. Yet none truly own their trade. Algorithms now play the role once held by charters and passes, deciding what is seen, sold, or silenced. The Company masked its rule through excellence in service; the platform perfects it.

Entrapment, too, has been redesigned. The logic that once governed bullion and trade now governs code and capital. The modern platform, like the Company before it, wins by loss. It spends not to sell, but to secure. Its goal is not margin, but monopoly.

In the seventeenth century, the Company paid in silver for silk and spices, a loss of precious metal justified by the long game of empire. Today's platforms repeat the ritual with capital. For years, they operate at staggering losses: subsidised rides, discounted products, free deliveries, zero-commission storefronts. Profit is deferred because control is compounding. Every discount buys a habit; every user becomes an unpaid participant in the system's expansion. Capital is deployed not to serve, but to ensnare.

India once celebrated the certainty of trade, unaware that certainty was the symptom of control. Predictability felt like progress: ships arrived on time, payments cleared, and markets appeared stable. Order seemed civilised, proof that partnership with the foreign trader had elevated commerce beyond chaos. Yet that very allure of stability was the first sign of surrender. When every route, every price, and every shipment became predictable, it was because they were already predetermined.

The same comfort now governs our digital age. Platforms, like the Company once did, offer frictionless certainty: instant delivery, seamless payment, reliable connection. The interface never falters; the algorithm never sleeps. But the ease we celebrate is the leash we do not see. The smoothness of service is not progress; it is permission, conditional, revocable, and designed to remind us of who owns the rails.

History's lesson is cruelly consistent: dependence always begins as convenience. The coming of the Company and, centuries later, the platforms are both stories of comfort that quietly replaced autonomy with dependence, what India once called trade, the world now calls technology. Why should both not end the same way?

3

FROM TRADE DEPENDENCY TO TECHNOLOGY DEPENDENCY

3.1 EMPIRE THROUGH DEPENDENCY

We the previous chapter, we saw how foreign companies arrived as partners, promising efficiency as progress and service as inclusion. Their charm was not conquest but convenience — ships once brought goods; apps now bring services. But every service that feels indispensable begins to extract something subtler than money: dependence. In this chapter, we follow that logic further. How does dependence begin, why is it seductive, and what happens when it matures into monopoly? Why does the origin of efficiency matter, whether it is foreign or domestic, when efficiency itself seems harmless? India has known this pattern before.

3.2 THE ARCHITECTURE OF ORDER

Technological innovation alone does not guarantee empire. Steam engines, ships, or silicon chips can enable power, but they do not assure it. What creates empire, as the following chapters will show, is the ability to turn invention into infrastructure, infrastructure into system, system into routine, and routine into the collective psyche.

A society that engineers efficiency into everyday life is bound, sooner or later, to become an empire.

Roads, ports, canals, and communication lines were not acts of invention alone; they were acts of organisation. They created nothing new — they merely reorganised movement. Efficiency, whether in motion, thought, or design, became a quiet form of dominance — a power exercised not through exerting more force but through reducing friction, increased coordination, and predictability.

Once such a path exists, few wish to haul goods by land or sail around Africa; everyone follows the shortest, smoothest line. Everyone seeks access to such efficiencies. The British, in turn, built on these efficiencies further. They created services around them that were continuous and predictable. Not just ships and railways, but timetables, market hours, price lists, banking systems, insurance, credit, and fixed interest rates.

Predictability became the stepping stone of progress: the ability to rely on an efficient system not once, but every time, whenever it was needed. When a service becomes deliberate, coordinated, and continuous rather than occasional or one-off, it transforms the chaos and uncertainty of daily life into certainty and order. The more predictable a service, the more society shapes itself around it. Human life naturally gravitates toward services that reduce variation, as if obeying a statistical instinct to minimise deviation, to smooth uncertainty, and to eliminate surprise.

Efficiency, thus, is not merely about speed but about reliability, the assurance that every link in a chain will hold. That is how a service becomes an institution and how infrastructure begins to govern life. Citizens come to depend on these systems, trusting them as constants in the background of daily existence.

For the British, this reliability, the regularity of services, became civilisation defining itself. The railway timetable symbolised order, a clockwork empire. The precision it demanded soon created the need to standardise time, measures, and exchange, so that the empire could move to a single rhythm. To make trains run on time, clocks across towns had to be synchronised, a quiet revolution that replaced local noon with Greenwich Mean Time. From this synchronisation arose the harder logic of uniformity. Time led the way, but soon the same precision extended to weights and measures (the Imperial System formalised by the Weights and Measures Act of 1824), postal rates (the Uniform Penny Post of 1840), currency (decimal reforms and fixed exchange standards), and telegraph codes (standardised in the 1850s for intercolonial communication). What began as a practical need for punctuality evolved into an ideology of precision, where every system had to move to a single, measurable rhythm.

3.3 PREDICTABILITY AS POWER

Predictable and standardised systems created predictable citizens. A worker who caught the 8 a.m. train every morning learned punctuality as instinct. Society itself became mechanised, its coordination outsourced to timetables. Punctuality, order, and foresight became signs of discipline and competence. Regularity matured the psyche of society itself; it created citizens who expected systems to work and believed that reliability was their right. What began as administrative precision evolved into social instinct, the confidence, and at times the arrogance, of expecting the world to run on time.

For the British, reliability, the regular rhythm of service, was civilisation made visible; it allowed its people to build lives around certainty, to take for granted what earlier generations and other nations could only hope for. Order gave the empire both economic edge and psychological authority. Once time, weight, distance, and currency were rendered into uniform units, governance itself could be expressed in numbers. Efficiency converted the subjective into the mathematical. It trained people to see regularity as virtue, variation as disorder, and flexibility as failure.

In India, such predictability was absent; Imagine a family setting out on a journey. Each journey required negotiation and contingency. A family wishing to travel might have to find a cart and a driver, bargain for fares, and then, upon reaching each stop, search for food and a place to sleep, all while hoping the road ahead was passable. In England, by contrast, one could simply consult the schedule printed in the newspapers, purchase a ticket at the station in the class of one's choice, and arrive on time, with porters ready to handle one's baggage. Where the English could plan, Indians had to improvised. Once societies begin to live along such services and depend on them, commerce, governance, and even imagination align to their rhythm. However, these systems are rarely possessions; they are shared services, not owned assets, yet whoever provides them quietly gains leverage over all who depend on them.

3.4 ROUTES OF POWER

Predictability itself is not the problem; every society seeks order. The question is who owns it. As mentioned, people naturally gravitate toward the predictable, building routines and expectations around what feels certain. Life is uncertain; predictability is a virtue. But when reliability and certainty is imported, it ceases to be virtue and becomes vulnerability. And when it is exported, it becomes power, the quiet ability to make others live by your clock, your standards, your sense of order.

The British understood this early. They did not invent India's trade; they only organised it. They brought schedules to ports, uniformity to paperwork, and discipline to delivery. In their hands, time itself became a tool of governance. When they built roads, ports, and railways, they were not only moving goods; they were synchronising an entire civilisation to their rhythm.

India found this rhythm irresistible. Ships arrived on time, goods moved swiftly, and contracts carried the comfort of certainty. This was service as seduction, as we saw in the last chapter. Dependence seemed profitable: trade expanded, margins improved, and British order appeared to multiply Indian prosperity. The empire's reliability felt like a gift, until it became the ground on which all commerce stood.

But with order came imbalance. Every ship, dock, and customs post became a British instrument. They built the infrastructure, and in building it, they built dependence. India came to rely on the very systems that displaced its autonomy. The more efficient the empire became, the more inefficient self-reliance seemed. The asymmetry deepened quietly.

First, the British did not simply move goods; they mapped them obsessively. Through revenue surveys, maritime charts, and the Great Trigonometrical Survey, India was redrawn into lines, ledgers, and coordinates. The empire learned to see the land in extractive terms, acreage and yield, ports and profit, labour and tax. It knew what each province produced, what each merchant exported, how much each region consumed, and which routes yielded the highest return. To trade with the British was to be seen. Long before monopolising trade, the empire saw the terrain in its entirety.

With that visibility came knowledge, and with knowledge, leverage. The Company now understood how to tilt advantage: lowering duties on its goods it preferred, raising them on others, and aligning taxes with its routes of profit.

With leverage came permission. Trade required the Company's approval, which goods could move, who could carry them, and at what cost. And with permission came preference: favourable rates and access for those who complied, and quiet penalties for those who did not. What began as preference soon hardened into favouritism, a quiet divide-and-rule of commerce. Those who traded on British terms prospered; those who resisted were quietly excluded.

Over time, favouritism turned into control. Weavers in Bengal were told what to weave; farmers in Bihar, what to sow. The empire's markets became India's instruction manual. Indigo replaced food crops; handlooms gave way to mills in Manchester. Prices were fixed abroad, enforced at home, and profits repatriated through shipping, insurance, and tax, all British-controlled.

And gradually, without much ado, control became submission. To trade with the British now meant acceptance, to live by the terms they set, the rules they wrote, and the values they defined. For merchants accustomed to and seduced by the new profits, this was the price of participation. The choice was not between freedom and dependence, but between inclusion and exclusion. A small class of traders and intermediaries, particularly from certain mercantile communities, prospered under this arrangement; their fortunes tied to imperial schedules and global markets. In pursuit of stability and gain, they accepted the empire's terms and, in doing so, allowed those terms to speak for the nation. Partnership by a few became permission for all.

And thus, what began as trade coordination and organisation ended as command. British management of routes and schedules became control of production and trade itself. Efficiency became instruction; reliability became rule. The British no longer facilitated India's commerce; they governed it.

Tariffs favoured British imports and penalised Indian exports. Local industries were taxed heavily or denied licences, while British goods entered duty-free. Indian weavers and ironworkers, once among the most skilled in the world, found their markets flooded with cheaper, machine-made products from Lancashire and Sheffield. Even essentials like salt, textiles, and grain were drawn into imperial regulation.

Every act of production or movement passed through British channels, ships, warehouses, insurances, and ports, each charging its toll. Indian merchants could not export without British brokers; Indian ships could not sail without British insurance. Even internal trade bowed to imperial rhythm: the railways carried British goods inland faster and cheaper than Indian goods could return to port. The empire did not need to block India's prosperity; it merely redirected it.

The Company no longer needed conquest; it had complete control of time, movement and trade. It commanded, dictated, and defined. What began as mastery over movement, routes and ledgers became a monopoly of trade.

3.5 THE MODERN MIRROR

The same seduction is alive today. The new empires arrive again with promises of efficiency, clouds instead of canals, platforms instead of ports. Their ships are silicon chips, their ledgers, algorithms. India once reorganised its economy around British schedules; today it organises its digital life around foreign servers.

The empire once promised "letters delivered anywhere in seven days"; the cloud now promises "uptime 99.999 percent." Dependence has migrated from infrastructure to interface, from trade routes to APIs. The promise is again convenience, and the price is again dependence. To revoke a licence or interrupt a cable is today's version of closing a port. The empire no longer needs ships; it controls the algorithmic switches. The monopoly is not yet visible, but it is already inevitable, written in every service we have come to trust.

And once again, India finds this rhythm irresistible. The same logic that once made the empire's railways seem modern now makes its platforms feel natural, even necessary. The country celebrates participation as progress, delighting in apps, messages, and endless reels, mistaking access for autonomy and partnership for power. But when reliability is imported, dependence is not far behind. Every shortcut still creates a chokepoint; every service still hides a gatekeeper.

3.6 RENTED SOVEREIGNTY AND THE NEW CHOKEPOINTS

Just as the British once controlled the movement of goods through ships, every byte that now crosses oceans travels through a handful of Western consortia, Google, Meta, Vodafone, and AT&T. The modern gatekeeper not only decides who may move but sees everything that moves, holding both the power of permission, to decide what data may pass through their networks, and the power of visibility, to know what passes, when, and between whom. Metadata has become the new customs ledger, recording every movement across the world's digital seas. The result is the same as before: participation depends on compliance; integration deepens dependence.

Such networks are no longer limited to cables beneath the sea; they are increasingly extending into the sky. Projects like Starlink aim to beam connectivity from space, expanding the same route ownership logic while bypassing governments and borders alike. Each new layer of infrastructure promises greater efficiency, predictability, and reach, even in the world's most remote regions. Yet what appears as universal access is still private architecture: another network owned elsewhere, operated elsewhere, and governed by terms technology-importing nations cannot negotiate.

When such power exists, it is rarely left unused. In moments of crisis, it becomes an instrument of leverage. The British once demonstrated the strategic power of route control through the Suez Canal. In both World Wars, they weaponised access, cutting off rival fleets and choking supplies. Today, the same principle endures, only the routes have changed. The exercise of such control over access reveals itself today through modern chokepoints.

The global financial network SWIFT is a case in point. When Russian banks were cut off from SWIFT in 2022, trade itself froze overnight, not by blockade or war, but by the quiet withdrawal of permission. The world discovered that a handful of servers in Belgium could halt the movement of trillions. This was the twenty-first-century Suez: whoever controls the route can still close the port.

Cloud computing follows the same logic. When Amazon Web Services suffered a global outage in 2025, payment gateways stalled, hospitals delayed records, and e-commerce froze from Singapore to São Paulo. The failure of a single cloud zone became a planetary pause, proof that the world has never been so efficiently fragile. Each such incident exposes the quiet leverage of the modern empire: the power to paralyse without confrontation, to discipline through dependency.

Governments now run on such leased infrastructure. Even when servers stand within national borders, through data localisation laws, true ownership often does not. What matters is not where data rests, but who can command it. The hardware may be local, yet the legal and operational control often remains foreign. This is dependency not of territory or proximity to data, but of technology. We still equate visibility with safety, as if being able to see the datacentre on our soil ensures control over it. It is a comforting illusion, born of an older world where possession meant power. In the digital age, ownership no longer follows sight.

Laws in the parent company's home country can reach across borders, compelling access to data regardless of where it is stored. Under the U.S. CLOUD Act, companies like Amazon, Microsoft, and Google must provide data to American authorities even when hosted abroad. China's National Intelligence Law (2017) and Cybersecurity Law (2016) impose similar obligations on Chinese firms. In this quiet contest of jurisdictions, geography no longer guarantees sovereignty; corporate allegiance decides whose command ultimately prevails.

Control also extends through operation, not just law. Even when foreign companies build data centres inside a country, they retain command over the protocols, maintenance, and system administration that keep those centres alive. Routine updates, software patches, and emergency responses can all be executed remotely, often by personnel thousands of miles away. The systems are too complex to be serviced locally; the know-how itself has become a form of dependency. The result is an invisible chain of control, a network that can be slowed, paused, or reconfigured at will. Analysts call this the "kill switch" problem: a foreign vendor's ability to interrupt or degrade essential digital services without crossing a single border. In extreme cases, it becomes an instrument of geopolitical leverage, the power to deny service without declaring war.

The deeper vulnerability lies in the hardware and firmware themselves. The servers, routers, and networking chips that form the cloud's backbone are largely designed and manufactured abroad. Every component, from the microcode that boots a machine to the software that orchestrates it, carries the potential for engineered flaws or backdoors. These may never be exploited, yet their mere possibility transforms technical design into political dependence. It is this anxiety that drives nations to pursue digital sovereignty: data-localisation mandates, indigenous cloud initiatives like MeghRaj, and regulations prioritising domestic oversight for critical systems.

Indian ministries store terabytes of data on AWS and Azure servers located within Indian territory yet governed by foreign law and operated through systems ultimately controlled abroad. This grants foreign entities near-total visibility into India's digital space. A policy shift or sanction elsewhere could ripple through India's networks, permissions revoked and access suspended, and the nation would have little recourse beyond diplomacy. The cloud has become both infrastructure and empire, a rented nervous system that technology dependent countries never fully own.

But beyond visibility and permission lies a deeper risk, path dependency. As technology advances, especially across frontiers like artificial intelligence, India is being drawn ever more tightly into the convenience of foreign systems. Even India's "Sovereign AI Mission," launched to assert independence, still leases compute time on NVIDIA clusters hosted on U.S. clouds. With technology evolving exponentially,

India's dependence on imported infrastructure grows with each generation, ensuring that its digital future remains authored elsewhere. Each new leap makes the escape harder, and technology migration prohibitively expensive.

This dependence is no longer theoretical. As seen earlier in the case of Microsoft's brief suspension of services to Nayara Energy, a single compliance action taken abroad was enough to paralyse systems central to India's energy infrastructure. The episode offered a glimpse of what is to follow, the quiet return of the same logic of pressure that once defined the British Empire. Then, it was tariffs and trade licences; today, it is platform access and compliance protocols. The forms have changed, but the function remains: to dictate terms without declaring authority. Dependence, in this architecture, is not enforced through conquest but through compliance, a preview of the likely deeper extraction to come.

The consequence of such dependencies is not merely vulnerability but influence. The priced extraction now is data. Information flows no longer just record a nation's activity; they predict it. And through such predictability, they shape markets, moods, and even the margins of political choice. The empire no longer needs to threaten or withhold access to technology services; it gains far more by guiding desire than by denying service. The aim is not to stop a nation's internet, but to steer its imagination, to make dependence feel natural, even voluntary. Profit now lies not in domination, but in direction, in shaping how people think, choose, and believe. Why influence, not coercion, is the preferred form of control is explored later in this book.

3.7 CHOOSING INCONVENIENCE OVER DEPENDENCE

Other nations are wary of such dependence and have taken a different path, a path of inconvenience over vulnerability. China and Russia, having watched the Western consolidation of digital infrastructure, built self-contained ecosystems that prioritise sovereignty over convenience. In China, platforms such as Google, Facebook, YouTube, and Twitter were never permitted to operate freely. Their absence gave rise to domestic alternatives, Baidu, WeChat, Weibo, Alibaba, and ByteDance, all operating within a regulatory perimeter controlled by the state. Foreign platforms may accuse this of censorship, but it also ensures authorship: the data, algorithms, and standards that define Chinese digital life are domestically governed.

Russia followed a similar instinct, but with a different intent. After U.S. and EU sanctions in 2014, Moscow began localising data under its Sovereign Internet Law. Unlike India's mere insistence on "data residency," Russia coupled localisation with infrastructure control and regulatory power. All major foreign tech companies operating in Russia were required not only to store data locally but also to submit to state access protocols and domestic routing systems. Those who refused, like LinkedIn, were banned. Local alternatives, Yandex, VKontakte, and RuTube, filled the vacuum. These services may lack global reach, but they guarantee insulation, enough for Russian digital life to continue even if the external networks go dark.

China has gone further, extending its sovereignty into hardware. Beijing's industrial policy explicitly targets self-reliance in semiconductors and AI chips. Huawei's Ascend series, Alibaba's T-Head processors, and SMIC's chip manufacturing capacity now power much of the country's domestic cloud and AI infrastructure. In September 2025, China Unicom inaugurated a data centre in Qinghai powered primarily by Chinese-made AI chips, about 72 percent of its 23,000 processors were domestically produced. Export bans on high-end American chips have only accelerated this drive. Nvidia's CEO Jensen Huang himself admitted that U.S. restrictions which triggered such bans were a "failure," reducing Nvidia's Chinese market share from 95 percent to around 50 percent, as Chinese firms rapidly advanced their own designs.

China's and Russia's logic is simple but decisive. Choose inconvenience over dependence. India seems to have chosen the opposite. Both nations understood that digital sovereignty requires redundancy, the ability to survive disconnection. India, by contrast, continues to rent its nervous system from abroad, choosing participation and convenience over control and capability. The same Microsoft that was condemned for blocking Nayara Energy was later celebrated when it arrived promising data centres and investment. Nationalism, it seems, dissolves quickly when dependency is repackaged as growth.

One must ask, then, why China and Russia were willing to bear inefficiency, while India so readily trades control for convenience?

4

From COLONIAL MONOPOLY *to* DIGITAL MONOPOLY

4.1 MONOPOLY AS PROCESS, NOT EQUILIBRIUM

Monopoly is typically framed as an outcome: the end result of firm growth, market dominance, and competitive suppression. This framing treats monopoly as a condition, a stable state arising from market failure and an equilibrium reached once markets stabilise.

This chapter approaches monopoly differently. It treats monopoly not as an equilibrium condition, but as a governing logic that shapes how power thinks and acts. Less an eventuality than a recurring disposition to organise and press advantage, monopoly persists as a programmable logic of control even as forms and technologies change.

Seen this way, monopoly is a governance style designed to learn, scale, and repeat. The East India Company operated through repeatable rules, feedback loops, and permissions that could be copied across colonie. Each success became the template for the next iteration, until domination itself was routinised through law, revenue systems, and administrative habit.

That playbook defined thresholds for action: when to trade and when to tax, when to discount to gain leverage, and when partnership should harden into possession. These were not spontaneous choices but calibrated strategies, refined into a process that could be replicated anywhere.

Modern digital monopolies inherit this same recursive architecture. Platforms prioritise penetration over profit, algorithms optimise relentlessly for scale, feedback systems distribute lessons learned across users, and rules continuously rewrite themselves to preserve advantage. What appears disruptive today follows the logic of monopoly that empire perfected long ago.

This distinction matters because monopoly reveals itself most clearly not in moments of abundance, but under pressure. When easy profits disappear, monopoly does not dissolve—it evolves and adapts. The British experience in India offers a clear illustration of this shift.

4.2 WHEN TRADE PROFITS VANISH

As shown in earlier chapters, by the mid-nineteenth century the British had mastered the mechanics of movement. Their ships, rails, and telegraphs bound the world to their rhythm.

In India, these instruments of control spread inland from the coasts, carrying the empire's order from port to province. The seas were charted, the ports secured, and time itself synchronised to imperial clocks. Trade routes were disciplined into obedience; merchants, once free navigators of exchange, now moved to the cadence of imperial contracts.

Trade had made Britain rich. But by the early nineteenth century, the Company's trade monopolies were no longer enough. Easy money from textiles, spices, and export levies had peaked. Competing European powers were encroaching, and Indian merchants increasingly operated beyond the Company's protection. To sustain extraction, Britain needed not new routes, but deeper leverage. Through the systems already in place, the logic of trade monopoly evolved into civilisational extraction.

Until this point, the life of the ordinary Indian had remained largely untouched. Beyond the presidencies, villages were still self-sufficient, growing what they ate and weaving what they wore. Local hierarchies governed land and labour; the rhythm of life followed the monsoon, not the empire's clock. The British had conquered Indian trade, but not yet the Indian way of life.

4.3 FROM MONOPOLISING TRADE TO LIFE ITSELF

To sustain its hunger, the empire turned its gaze inland, towards land, labour, and life itself. Land revenue from India would offer what trade no longer could: a steady and expandable source of income. To collect rent at every point of survival, Britain had to eliminate every system that operated outside its control.

Commerce alone would no longer suffice; the new frontier was the field, the well, the loom, and soon, the classroom. What began as profit through trade would evolve into permanence through civilisational redesign and rent collection. A monopoly on how life itself could be lived.

By the mid-nineteenth century this shift was complete. The Company's commercial ambitions had matured into a machinery of rule. Taxation, land, labour, and law were drawn into a single administrative rhythm. Railways, telegraphs, and revenue settlements tied distant provinces to the same pulse, turning exchange into extraction. Profit no longer came from trade participation, but from interference, from restructuring how Indians worked, produced, and lived.

The chapters that follow trace how this shift was orchestrated. The starting principle was elimination. The British did not seek to compete with India's systems of life; they sought to erase them. Every craft, crop, and custom that offered self-reliance was seen as a threat to imperial order. The aim was not merely to rule, but to render India incapable of existing without the empire's hand. The killing of capability was the governing objective.

4.4 KILLING CAPABILITY

Monopoly must first destroy its competitors' capability before it can rule. Conquest may seize territory, but monopoly disables the capacity to create. In the colonial context, monopoly was therefore not merely economic; it was civilisational. It operated on two levels. The first concerned physical capability: the power to determine what could exist, be produced, or be known. The second concerned psychological capability: the power to shape what could be imagined. When authority dictates not only what can be made but what can be conceived, control shifts from external force to internalised belief.

Britain did not simply rule India; it redesigned it to prevent recovery. India's defeat was not the loss of armies but of capability. A nation can recover from invasion, it cannot easily recover from the systematic erosion of its capacity to make, decide, and dream. By severing the link between work and survival, the British ensured India could not exist outside their economy.

4.5 THE FIVE E'S OF EXTRACTION

This destruction of capability was not random. It followed a pattern. From cotton to code, from kings to corporations, the logic is identical: to monopolise the future, empire must first destroy all competing potential, and with it, the possibility of alternatives. Five recurring strategies reveal how capability was killed and captivity made permanent. Instances and episodes that correspond to these five strategies are well known to history; my purpose here is not to recount them, but to place them within a single interpretive frame, a design through which empire perfected the art of monopolising conditions of civilisation, and whose pattern may help foresee its modern reincarnations.

E1 — Erase Production

The first act of the empire's monopoly was to dismantle local industry. Before British rule, India's economy was a dense mosaic of craftsmanship and trade, with cotton from Bengal, silk from Murshidabad, and steel from Mysore. Production was decentralised but disciplined, sustained by inland markets and vast internal demand. To rule India, the British first had to unmake this self-sufficient order.

Factories were not burned but legislated out of relevance. Trade laws and tariffs rewired the flow of value until production no longer served India but Britain. Some industries were taxed, underpriced, and replaced by machine-made imports; others were shut down under the pretext of "safety" or "environmental conservation," their furnaces dismantled, and licences withheld.

Indian industry, once the world's benchmark, was driven to extinction. The same goods once crafted in India were now imported at high prices and duty-free, while the raw materials to make them were exported to Britain.

The British also de-skilled India. Guilds that had governed wages and workmanship were dissolved and replaced by so-called "free" labour markets policed by British magistrates. Tax codes reclassified craftsmen as agricultural dependents, erasing their identity as skilled producers. Under colonial rule, ustads and karigars alike were reclassified as labourers, men of mastery reduced to men of maintenance.

The Workmen's Breach of Contract Act of 1859 bound such skilled workers to employers by law. Any labourer who left before a contract's end could be arrested and jailed. Many were trapped in mismatched or menial work, such as a weaver digging ditches or a blacksmith hauling grain, unable to return to their craft without breaking the law. Their choice was to abandon their skill or starve. Between self-erasure and starvation, most chose survival.

E2 — Enclose Resources

Once India's industries were broken, the next step was to control what sustained them: land, water, forests, even the seasons. Cultivators who had once shared harvest risk now owed fixed rent to the Company, unchanged even when the monsoon failed. Land was measured not by fertility but by fiscal potential. Each acre became collateral, each farmer a debtor. The purpose of land was not food; it was revenue.

Once land could be taxed, it could be instructed. Grain gave way to cash crops. Cotton in Bombay, indigo in Bihar, opium in Bengal, wheat and sugarcane elsewhere. Under the indigo system, European planters lent money that bound peasants to grow a crop they could neither eat nor price. They were fined for growing anything else. The fields still bloomed, but not with food for Indians. Between 1876 and 1878, millions starved in Madras and Mysore while grain exports to Britain hit record highs. Indians died of starvation in the shadow of full granaries.

Even forests were enclosed. The Indian Forest Acts of 1865 and 1878 claimed what had once been shared. Within three decades, over a quarter of India's forests passed into Crown control. Villagers who grazed cattle or gathered wood became trespassers on their own land. To walk in a forest now required a permit. The pretext was conservation; the purpose was timber, for ships, for railways, for empire.

From forest to sea, nothing remained free. Even salt, the most common and necessary of elements, was taxed. There was no pretext this time, no veil of reason, only profit. This revealed the true morality of empire: to create a monopoly not just of trade, but of life itself.

By the century's end, the British had achieved an ecological monopoly. India's air, soil, and seed were folded into imperial arithmetic. What had once been a living web of interdependence was redrawn as a grid of taxable assets. Nature itself had been claimed, nationalised and taxed.

E3 — Enforce Terms

Exploitation begins when negotiation ends. The empire institutionalised one-sided exchange and made it appear lawful. Consent, once the basis of trade, was replaced by compliance. Taxation became a daily ritual of subordination, repeated every harvest, every wage, every life.

The Permanent Settlement of 1793 transformed taxation into certainty. Designed by Cornwallis to secure the Company's income, it fixed revenue not by yield but by imperial expectation. Even when crops failed, payment was still due. Risk flowed downward: zamindars lost estates, peasants lost homes. Within a decade, a third of Bengal's land had changed hands. What had once been shared obligation became a hierarchy of debt.

When payment became impossible, people turned to moneylenders. Interest soared, and default meant dispossession. The Deccan Riots of 1875, in which peasants attacked lenders' ledgers, were a revolt against paperwork, not people. Britain's answer, the Deccan Agriculturists' Relief Act of 1879, claimed to protect debtors but instead codified their bondage. Private loans became state-supervised dependence; extraction gained a legal face.

To plunder with a clock of legitimacy, the colonial state extended formal legal access. Yet, as noted by several historians, this was largely performative, a theatre of rights that masked the machinery of extraction. The Evidence Act and the Civil Procedure Code replaced Indian custom with

British codification, ensuring that every dispute, over land, debt, or water, was judged in English and settled on imperial terms. Courts turned justice into revenue; filing fees and fines replenished the very coffers in dispute. Delays, adjournments, and procedural backlogs became instruments of bureaucracy, denying justice but keeping its illusion alive. Legal battles soon cost more than the lands or claims they sought to defend. And so, under the guise of law, untold tracts of land passed from the poor to the powerful.

By century's end, labourers lived in a closed loop: borrow to pay tax, work to pay interest, die to clear arrears. The British had created the very poverty they later claimed to civilise, a loop designed to extract rent indefinitely. As R. C. Dutt wrote, the Indian cultivator paid for his chains in advance. There was no bargaining, no negotiation, only terms enforced.

E4 — Exterminate Alternatives

Monopoly matures when another's success itself becomes punishable. Once India's industries and land were subdued, the empire moved to eliminate every remaining source of self-rule: rulers who could govern, merchants who could fund, financiers who could sustain. What could not be conquered was converted; what could not be converted was crushed.

The Doctrine of Lapse (1848) annexed states without heirs, conquest without war. Jhansi, Satara, and Nagpur fell; those spared were bound by "subsidiary alliances," forced to fund British troops and surrender diplomacy. After 1857, princes were restored only as ornaments. "We have given them the shadow," said Viceroy Lytton, "and kept the substance."

With power domesticated, finance followed. The abolition of Indian mints in 1835 forced all coinage into the imperial rupee, collapsing regional monetary systems. The Presidency Banks monopolised currency issuance and government accounts, dismissing Indian financiers as "inexperienced." Commerce was fenced in by the Companies Act of 1866 and the Marine Insurance Act, which barred Indian firms from large-scale trade. Merchants who once financed voyages from Basra to Batavia were reduced to brokers for London, becoming middlemen in their own markets.

By century's end, India's political and economic autonomy had been extinguished. Rulers governed at Britain's pleasure, merchants traded on British terms, and finance flowed through British banks. The empire had not only conquered India's wealth; it had abolished her right to create it.

E5 — Empty the Imagination

The final conquest was not of land or labour but of the mind. Once India's ability to produce, trade, and govern was dismantled, the empire moved to control meaning itself: how people thought, learned, and dreamed. Macaulay's Minute on Education (1835) dismissed India's classical learning as inferior and replaced Sanskrit and Persian with English instruction. This was not education but engineering. To think in English became intelligence; to think in one's own tongue became backwardness. Over time, imitation replaced imagination. Obedience was recast as virtue, acceptance as reason, and faith as superstition. Indians began to see themselves through imperial eyes.

Control of thought demanded control of information and speech. The press, like the classroom, became an instrument of calibration. To speak freely, one first had to speak acceptably. The Vernacular Press Act (1878) criminalised "disaffection," silencing regional voices while English papers read by officials remained untouched. What could be said, and what could even be imagined, was policed with care. Those who defied it were censored, fined, or jailed. Letters and telegraphs, once symbols of connection, became channels of surveillance, opened, read, and inspected in the name of security. Communication, once a right, became a privilege measured by loyalty.

By century's end, Britain had achieved what conquest alone never could: a monopoly over imagination itself. The ability to envision a different order had dimmed. Permission replaced pride, and acceptance seemed wiser than courage. The Indian spirit itself was monopolised.

4.6 WHAT WAS LOST CANNOT BE CALCULATED

When monopoly matured, extraction became invisible, enforced through taxes and tribute, and guaranteed by obedience and a controlled imagination. The economic drain of colonial rule has been documented in detail by scholars from Dadabhai Naoroji to Shashi Tharoor, a plunder now estimated at over thirty-five trillion dollars in today's value. Their work shows how trade became tribute and profit turned into rent. There is little to add to that record of loss.

What remains incalculable are the opportunity costs and the psychological costs they conceal. The futures that never unfolded, the prosperity that might have been had Britain not looted but used the gift of technology to enlarge the pie rather than steal from others. And the psychological damage that endures: a civilisation taught to doubt its own worth, its own mind, its own possibility. The loss was not only of wealth but of will. The quiet erosion of confidence that limits imagination even today.

What matters now is to see how the same pattern of monopoly endures in new form: not through the export of gold or cotton, but through the extraction of data, attention, and insight. The invisible resources of the twenty-first century.

As reasoned in this chapter, monopoly survives not by force, but by learning faster than its environment, converting resistance into information and adaptation into design. Every challenge becomes feedback; every failure, a prototype. What appears as disruptive is in fact an evolving system that studies its surroundings, anticipates threats, and rewrites its own rules to preserve asymmetry.

In this sense, monopoly behaves less like a state of being and more like a process. A self-updating form of power, where power is performed rather than possessed. What the Company once achieved through policy and paperwork, algorithms now pursue through prediction and pattern recognition. The code has changed, but the logic remains.

4.7 MODERN MIRROR

As we can see, the British were remarkably busy in the first half of the nineteenth century, drafting laws, designing systems, and building institutions that taxed, measured, and managed nearly every aspect of Indian life. It was a short but decisive era, when Britain quietly shifted gears from trader to coloniser. Yet this phase is rarely discussed. We remember the trading Company of the eighteenth century and the exploitative Raj of the late nineteenth, but the critical interlude between them, when control was coded into law, finance, and education, often slips through the cracks of public memory. In hindsight, it was the most creative moment of empire: a blueprint of control so intricate that its purpose was invisible to those living through it.

Only now, with distance, can we see how these scattered acts, in agriculture, banking, schooling, and speech, formed a single design. Today, we stand at a similar threshold. Algorithms and technology platforms are still in their trading phase, offering service and convenience. But will they, too, one day change gears, from providers to governors? The signs are emerging, faint but familiar. As before, each regulation, update, or policy may appear isolated, even benign. Without knowing the end game, it is hard to connect the dots. Yet vigilance begins in recognising patterns, and history teaches us that hegemonic powers rarely announce their intentions. The pages that follow explore these modern parallels, imperfect, perhaps premature, but necessary while there is still time to see the pattern forming.

4.8 THE AUTOMATION OF THE MONOPOLISATION PROCESS

In our age, global platforms are still in their trading years. Profit is not yet the goal; penetration is. Just as the British once paid more to enter a market than they earned from it, today's technology giants pour billions into free services, subsidies, and loss-making expansion, not to make money now, but to ensure that no one else can later. They are still merchants, not monarchs. The real prize is ubiquity: to be the channel through which everything flows. Once a platform owns that channel, of communication, commerce, or cognition, profit becomes inevitable. Until then, expansion is survival.

This is why most digital empires remain unprofitable or marginally so. Cloud infrastructure, AI, logistics, streaming, these are today's canals and ports, built at extraordinary cost because their builders believe the future will run through them. Profit will come later, once the flow itself is captured. In this sense, we are still in the age of convenience, not control.

Technology is advancing too quickly for saturation. Each leap, from social to mobile, from cloud to AI, opens a new frontier, another ocean to cross. For today's tech corporations, there is still always a new domain to cross, another continent of data to discover. Yet history suggests this phase will not last forever. When technological expansion slows or when enough advantage has been secured, when everyone is connected, every habit mapped, every market linked, the logic of penetration will give way to the logic of extraction.

That is when empires, imperial or digital, shift gears. When they can no longer grow by serving, they begin to grow by governing. We are not there yet, but the outlines are forming. The question is whether, when that moment arrives, we will recognise the shift, or once again mistake the machinery of control for the machinery of progress.

As seen historically, monopoly begins by hollowing out independent creation whatever proves successful. Today, platforms that once connected buyers and sellers are starting to quietly absorb both. E-commerce giants first hosted small merchants; now they use sales data to identify best-selling products, launch near-identical versions under their own private labels, undercut prices, and dominate logistics, hollowing out the very ecosystem that made them successful. The marketplace no longer rewards entrepreneurship; it welcomes it, studies it, copies it, and eliminates it. The true commodity is not the product sold, but the formula of success itself.

This logic now extends far beyond online retail. Cloud providers and app platforms follow the same learn-and-replace pattern, studying which services thrive on their infrastructure and then releasing built-in versions that quietly displace them. Third-party software that once filled essential gaps is re-engineered into "native" features, turning the host into both marketplace and competitor. Even in everyday tools, the shift is visible: fitness and wellness apps that once flourished independently were soon mimicked by built-in trackers bundled with devices, erasing an entire ecosystem of innovation. Hosting ends as hijacking.

Delivery platforms, too, began as logistics intermediaries, connecting local restaurants and grocery stores to customers. Over time, they studied what sold best, at what time, and in which neighbourhoods. The same data that once powered convenience became competitive intelligence. Using it, the platforms began opening their own "dark kitchens" and fulfilment hubs, invisible businesses with no storefronts, producing algorithmically optimised meals and goods designed to outperform their clients on the very apps meant to serve them. What started as partnership evolved into parasitism. The restaurant and retailer were reduced to data sources; the platform became both distributor and producer.

4.9 AI AND THE IRRELEVANCE OF COGNITIVE LABOUR

Once again, the production of value is under threat, this time from artificial intelligence. Just as the British once made India's weavers and ironmongers redundant, AI risks making India's graduates redundant, not by denying them work but by making their work irrelevant. The skills that once sustained employment - coding, accounting, customer service, report writing - are being absorbed into algorithms that perform them faster, cheaper, and without fatigue. What was once industrial displacement has become intellectual displacement.

For a nation whose economic rise was built on exporting educated professionals, programmers, business executives, analysts, accountants, and customer support professionals, the danger is existential. AI does not need their participation to produce value; it only needs their data to perfect itself. Each prompt, ticket, and line of code becomes training material for the very system that replaces its creator. When a civilisation's most employable talent, by volume of graduates, becomes surplus to the process of value creation, it faces not unemployment but obsolescence. A quiet echo of the deindustrialisation that once dismantled India's craft economy.

India must ask what these millions of students will do. We must pause and think about the consequences of doing nothing. To see what may come, we only need to look back and ask: what happened to the skilled craftsmen of yesterday?

They were forced into the labour of their age. Tomorrow's unemployed youth may or may not be breaking big rocks into smaller rocks, but it will be something equivalent by the standards of our time, though we cannot yet imagine it. I doubt the ironmonger of Mysore could have foreseen the day he would be quarrying stone for a road he once forged tools to build. The idea now is to recognise the threat early, to use history as a marker of what is to come. Nobel laureates and leading economists are already warning of this as a new civilisational risk—the erosion of meaningful human capability itself. India, positioned at the lower end of the technology spectrum, with millions of graduates and a vast labour market celebrated as a demographic dividend, may in fact face the greatest reversal. What is now hailed as a great abundance could, without foresight, become a great burden.

The other side of de-skilling is dependency. What automation once did to muscle, AI now does to mind. Each task it performs removes the need to practise it. At first, this feels liberating and effortless: accounts balanced faster without accountants, reports written better without executives, code generated instantly without programmers. But with every convenience, a capability is quietly lost while a dependency is slowly built. As AI absorbs repetitive and procedural work, it erases the very practice through which skill was sustained. Within a generation, professions that today use AI as assistance may depend on it for cognition itself.

The empire no longer needs to ban learning; it simply replaces the need to learn. And once control over these professions rests in the hands of foreign corporations, the new raj will not rule land or labour, but logic itself. And like before, it will decide how accounts are to be balanced, what reports are allowed to state, and how much tax is to be collected. Any calculation outside its framework will be branded as error, illegal, incorrect, or simply invisible. The logic of compliance will once again be coded into the system itself.

As history shows, what begins as enablement often ends as dependence. The British taxed India's weavers until the loom went silent; today, algorithms promise convenience until thinking feels redundant. When systems begin to decide, remember, and recommend on our behalf, the ability to think, create, and imagine for ourselves quietly withers. What was once knowledge becomes a service; what was once culture becomes content. The more a society outsources its judgment to foreign platforms and algorithms, the less it remembers how to make meaning for itself.

4.10 PARTICIPATION WITHOUT POSSESSION

India again stands at a familiar threshold. The country does not own the platforms on which its creativity runs, nor the models that interpret its data. Local talent codes, but the code itself trains foreign intelligence. As once with cotton and steel, the raw material of data, skill, and human intent is exported, while the finished product, AI capability, returns as service.

Russia and China resisted the import of such foreign design by building their own platforms and closing their digital borders. India remains open, confident that participation equals progress. Yet because India never truly owned its digital resources, it lives under the illusion that there is nothing to be stolen. Unlike the age of cotton or steel, there is nothing to seize because everything is given. The platforms, the cloud, and the search engines all arrived pre-built and pre-owned. And because there is nothing to steal, India remains blind to what is being taken.

Where China fenced off its digital territory by banning global search engines and cultivating its own ecosystems of search, commerce, and artificial intelligence, India opened its gates in the name of digitisation. The result was immediate participation, but no possession: a digitised India made by the West, not made in India.

Today, China uses its platforms to shape citizen behaviour and steer national outcomes. Digital India, by contrast, has no such power; it lives within platforms it does not own and policies it does not write. The consequences of not owning one's digital infrastructure are now unmistakable: once digital participation depends on foreign systems, the power to shape society, and to set the terms of its future, quietly slips into foreign hands.

Cloud providers control uptime, pricing, and scale. App stores gatekeep distribution through proprietary standards. Advertising networks ration visibility across the web. Data itself, the gold ore of the twenty-first century, lies locked in servers beyond India's control, even when those servers stand on Indian soil. Those servers train algorithms that learn from and shape the Indian user, market, and mind — the true gold of the digital age.

India's political, bureaucratic, and corporate communications all transit through infrastructures beyond its jurisdiction. The idea of Trojan horses hidden within such systems is no longer fiction but a known possibility. Believing that these communications are fully secure, encrypted, and unread is, at best, a comforting illusion. In a country like India, where majority of the digital infrastructure comes from abroad, the only true way to remain invisible to foreign agencies is to abstain entirely. A difficult thing to do. Yet this caution is echoed in practice by India's National Security Advisor, Ajit Doval, who has publicly stated he does not use a mobile phone or conventional digital communication, warning that dependence on foreign technology platforms poses a grave national security risk.

4.11 FROM CONVENIENCE TO COMPLIANCE

Negotiation, once the foundation of all trade, is already fading. For the individual user, there is little to bargain with: every term, response, and appeal is automated. The screen has become both marketplace and magistrate; one can only accept, appeal, wait, or leave. Even corporations now find themselves at a disadvantage, as access to essential services can be throttled or withdrawn, forcing compliance rather than consent.

Once monopoly is secured, convenience turns to control. The same platforms that once subsidised entry begin to tax participation, no longer competing for users but commanding them. Like before, users can only adapt to survive.

Early signs of this quiet conversion from empowerment to compliance are already visible across the digital economy. Search results shift without warning, leading to abrupt collapses in customer reach and forcing firms to adapt to new, opaque requirements. Price structures change overnight, unsettling entire business models built on borrowed platforms. Each adjustment deepens the dependence, binding businesses ever more tightly to the compliance regimes of foreign systems.

Businesses built entirely online find themselves paying ever more invisible tributes for visibility, for ranking, and for survival, without the option to negotiate. Freelance platforms deduct multiple layers of fees for service, payment, and promotion without a single round of discussion. Gig drivers and food delivery partners face fluctuating payouts determined by code they cannot question. Influencers and content creators, too, must conform to Western algorithms, aesthetics, and appetites, tailoring their expression to trends they do not set, audiences they do not own, and rules increasingly policed by the very platforms they depend on.

When monopoly matures in today's digital era, exterminating alternatives takes a subtler form. They are no longer banned or destroyed; they are absorbed or outdesigned. Platforms that once claimed to democratise access now quietly fence it. Mobility platforms promote their own fleets or preferred partners in search results, while restricting access to route and pricing data that independent competitors need to survive. Accommodation and delivery platforms discourage cross-listing, rewarding exclusivity through better rankings, lower commissions, and promotional visibility. App stores enforce proprietary payment systems, taking a share of every transaction while blocking third-party billing. The rules of participation have become the architecture of exclusion. Access is costly but comes with temporary privileges such as visibility, reach, and relevance that bind participants more tightly to the system granting them. Those who refuse lose relevance in the digital world.

Once work, access, and value are enclosed, monopoly moves to the final frontier, emptying the imagination itself. The goal is no longer to own production, but to own perception, to decide what people see, seek, and strive to create. Algorithms now shape desire more efficiently than any decree ever could. They reward imitation, not originality, outrage, not reflection. One steers retail consumption; the other steers political discourse. Creativity becomes predictable, measurable, and optimised for engagement. The result is a quiet narrowing of human possibility. A civilisation taught to think in templates, designed once again by the West.

Just as earlier imports of Western technology first arrived as trade, then created dependencies that evolved into monopolies extracting resources and shaping thought, today's technologies appear to be following the same arc. As yesterday's companies and today's platforms are both profit-driven, it is therefore reasonable to hypothesise that they may eventually evolve toward a similar reach and logic of control.

The precise form of how India's technological dependency will translate into monopoly and extraction is yet to emerge and remains uncertain. It will depend on how technology advances, and on how creatively the West, which still leads in design, infrastructure, and capital, chooses to extract and govern value. What we see today are early signals, not outcomes: tendencies in pricing, access, and control that could solidify into a new structure of governance if left unchecked.

The purpose here is not to forecast that outcome, but to recognise the process already unfolding and the direction in which the present is moving, while there is still time to respond.

5

From POPULATION GOVERNANCE *to* BEHAVIOURAL ANALYTICS

5.1 THE CALCULUS OF SOCIETY

The previous chapters traced how empire secured control through trade, monopoly, law, and the systematic destruction of indigenous capability. Yet material domination alone cannot explain how such control endured. It required a mechanism of population control. That mechanism was rooted in a governance tradition that long predated the Raj and depended on a science capable of classifying, measuring, and managing people.

Examination of this tradition, the belief that society could be measured, quantified, and engineered, remains strangely absent from Indian scholarship. Macaulay appears prominently in the Indian memory of cultural domination and is often treated as its architect. In reality, he was merely the final emissary of this lineage, the Indian-facing expression of a two-century-old project to study, classify, and reshape human behaviour.

This chapter follows that lineage back to its origin, asking where this science began, why it was needed, and what conception of human behaviour it was designed to shape. To answer this, we return to the birth of the idea of social manipulation, an origin seldom explored, yet foundational to the empire's machinery and to the methods of influence that endure today.

Long before India encountered English education or British newspapers, the British had already developed a theory of human predictability. They called it the physics of society, political arithmetic, an early science of governing the mind. But the ambition was unmistakably modern: to quantify people, to classify them, to rank them, to anticipate their behaviour, and ultimately to engineer it. What we now call "behavioural analytics," they once called the calculus of society.

Unlike Indians, who readily celebrate their ancestral achievements, Britain left this intellectual lineage obscure, then and now. It had no reason to draw attention to it. These ideas were not designed to inspire pride or spark cultural reverence; they were designed to organise power. Britain highlights the romance of Shakespeare's English, Darwin's study of man, the rise of rational economics, and the gift of the steam engine, but not Petty's political arithmetic or Buckle's social physics, the intellectual tradition that sought to turn society into something measurable and governable. These were revolutionary innovations in their own right, developed meticulously but without fanfare. By the time the British arrived in India, they brought with them not merely ships, guns, and railways, but a two-century-old behavioural science waiting for a laboratory.

5.2 THE ORIGINS OF POPULATION CONTROL

In seventeenth-century England, the grip of social order was loosening. The divine right to rule was no longer accepted. Parliament had emerged as a counterweight to the crown, granting ordinary men political claims that would have been unthinkable a century earlier. The tension between royal authority and new aspiration tore the country apart, culminating in the English Civil War and the unprecedented spectacle of a king being led to the scaffold.

The execution of Charles I shattered the foundations of governance. The monarchy had justified its power through divine sanction; the Church had enforced it through moral authority. With both shaken, England found itself without a coherent organising principle. The old certainties of kings, God, and hierarchy could no longer command universal obedience. The common citizen, exhausted by war, lived in a society stripped of its traditional anchors, a world suddenly unsafe, unstable, and adrift. What emerged was anarchy, and anarchy was unbearable.

It was in this climate of confusion and fear that Thomas Hobbes wrote Leviathan. He was not merely responding to political turmoil but attempting something radical: to deduce, through logic and reason, the principles by which human beings could coexist without tearing one another apart. Hobbes sought to construct a science of human interaction, an account of behaviour, fear, cooperation, and authority grounded not in divinity but in a mechanical understanding of human nature.

Mechanistic science, at that time, was transforming Europe's understanding of the natural world, replacing mysticism with mechanism. Hobbes believed the same logic could be applied to society. Hobbes imagined that people could be treated as "matter in motion" and understood through their predictable impulses of appetites and aversions.

Each person, Hobbes argued, pursued their own motion, their desires, until colliding with the desires of others. Most people had moderate wants, but a few possessed boundless appetites, and it was they who produced conflict, disorder, and civil war. Stability therefore required the many to surrender part of their freedom to a central authority strong enough to restrain the few. Only such a "Mortal God" could prevent society from sliding back into anarchy.

In effect, Hobbes repackaged the old logic of rule, the need for a sovereign, through the language of physics. Obedience, in this view, was not a moral duty ordained by divine authority but a behavioural outcome that could be channelled and stabilised. Yet, while the conclusion was old, his approach was novel. Instead of looking upward to those who ruled, he looked downward at those who were to be ruled and arrived at the same conclusion about civil obedience. In doing so, he revealed that the true enemy was not tyranny, but anarchy.

It is in this sense that Hobbes inverted the traditional problem of governance. Instead of asking who should rule, he asked who was to be ruled. Instead of studying sovereignty from above, he studied behaviour from below. This shift — looking downward, not upward — marked the intellectual birth of social engineering. If human nature could be described in mechanical terms, then society could be organised through mechanical principles. Hobbes' logic was swiftly absorbed into domestic governance, providing intellectual cover for stronger central authority, tighter control of dissent, and the expansion of policing after the Civil War.

And thus, in the shadows of civil war and from the fear of a society without order, emerged a radical idea. Instead of looking upward at the sovereign, Hobbes looked downward at those who were to be ruled, treating their behaviour itself as the proper object of governance. If conduct was driven by appetites and aversions, then interactions could be shaped through incentives, fears, and constraints. Obedience was not divine duty but behaviour that had to be conditioned and stabilised. This downward shift in perspective laid the intellectual groundwork for Britain's later sciences of classification, measurement, and control.

5.3 POLITICAL ARITHMETIC AND THE BIRTH OF STATISTICS

If Hobbes supplied the philosophy for managing human behaviour, Petty laid the foundations of the arithmetic that made such management possible. Through what he called "political arithmetic," Petty argued that society could be understood only to the extent that it could be measured: births, deaths, wealth, labour, disease, productivity. By gathering as much information as possible, he believed governance could be reduced to calculation. He translated the Hobbesian need for obedience into a counting problem: numbers to be gathered, tables to be built, and lives to be recorded and arranged. Petty's political arithmetic measured English mortality, labour, poverty, and wages, turning the daily life of Britons into tables of productivity, dependence, and risk. These findings shaped life insurance schemes, public health planning, and calculations of military manpower, embedding quantification into everyday governance.

Petty was a founding member of the Royal Society. His arithmetic of governance quickly found intellectual allies who shared his conviction that society could be understood through numbers. What followed was a wave of counting, tabulating, and census-making — taken up by figures as diverse as Halley, Laplace, Fourier, and Nightingale. Soon these scholars began to notice something extraordinary: beneath the apparent chaos of human life lay patterns of almost mathematical regularity. Süssmilch, for instance, observed that male and female births balanced with uncanny precision year after year. From such regularities, philosophers like Kant concluded that society obeyed hidden universal laws.

The discovery of such patterns pushed scholars to search for deeper laws of social life. They noticed that much of this data about social life, wages, productivity, mortality, and other human measures, fell into predictable shapes. Values clustered around a central peak and declined symmetrically toward the edges, forming the bell-shaped curve that would later be known as the normal distribution.

To make sense of these shapes, the field developed the tools of averages and deviations, measures of how far any one person stood from the centre of the group. But knowing how far was only the beginning; the next question was how likely, how probable, any deviation might be. This gave rise to probabilities and error curves, mathematical devices that estimated the chances of an individual's departure from the mean. And from this emerged the idea of predictability itself, the power to foresee what would happen by knowing the typical values of the past and the expected deviations from them.

Indeed, the very name "statistics" arose from this impulse. Gottfried Achenwall coined the term Statistik in 1749 to describe the science of the natural states of society — a discipline meant not merely to count people, but to understand how a population functioned as a system.

These tools soon reshaped how people understood the world. Averages made it possible to speak of the "typical" wage or the "expected" lifespan; deviations showed how far any one person strayed from that norm — the labourer earning half the ordinary wage, the shopkeeper demanding three times the usual price. Probabilities allowed scholars to estimate the chances of events once thought chaotic — the likelihood of falling ill, of surviving an epidemic, of earning above or below the mean. Predictability followed: governments could now estimate harvest yields, anticipate tax receipts, and forecast the burden of disease with a confidence unimaginable a century earlier. What began as simple counting evolved into a way of seeing society through patterns, not individuals.

5.4 FROM AVERAGES TO MORAL HIERARCHIES

Statistics, the science of averages and deviations, soon acquired a moral sheen. Adolphe Quetelet fused these strands into the idea of l'homme moyen — the Average Man — whose physical, moral, and behavioural traits he treated as society's natural centre. Deviation from this centre, though statistical in nature, quietly took on a moral tone: departures from the mean began to look like departures from order. In Victorian Britain, this logic hardened further — variation became defect, irregularity became vice, and statistical difference was recast as a sign of weakness or inferiority. The worship of measurement, ranking, and the normalisation of human beings began here.

Francis Galton reinterpreted Quetelet's deviations from the average as markers of innate superiority or inferiority. If society displayed bell-curve variations, Galton reasoned, then individuals could not possess equal value; some were inherently "better," others "worse." And if human beings did not share equal value in ability or character, then they could not be equal as social units — not equally capable of judgment, voting, or civic responsibility.

Armed with the mathematics of correlation and regression, tools he pioneered, Galton set out to classify individuals and populations according to innate worth. Human value could now be ranked from the "gifted" to the "feeble," from the "fit" to the "unfit." He argued that intelligence, talent, ability, achievement, and even character were unevenly distributed across society. His most audacious claim was that the most "desirable" qualities were hereditary, clustered within certain families, classes, and races.

What Quetelet had treated as natural statistical variation, Galton treated as hereditary destiny. This was the moment when descriptive statistics hardened into self-righteous ideology. The consequent leap into selective breeding was small. If desirable traits clustered in certain families and undesirable ones in others, then improving the distribution meant encouraging the "fit" to reproduce and discouraging the "unfit."

Galton's conviction that genetic inheritance determined human ability led him to develop the early tools of biometrics, the statistical study of biological variation, and made him one of the founding figures of modern eugenics. His successor, Karl Pearson, famous for formalising the correlation coefficient that bears his name, extended this project even further by embedding statistical tools and techniques within a broader programme of racial hierarchy, hereditary classification, and the defence of empire.

Galton's work did not stand alone. It grew within a broader British intellectual climate that sought not merely to measure society, but to rank it. Henry Thomas Buckle was instrumental in deciding what attributes would become the basis of this ranking. Crime rates, suicide rates, literacy levels, innovation records, "moral progress," even the supposed influence of climate and food were converted into statistical indicators of a nation's character.

Buckle believed these patterns displayed such regularity that he drew a sweeping conclusion: nations could be arranged on a ladder of "mental development," with some deemed "advanced," others "immature," some "progressive," and others "regressive." Once civilisation was framed as a hierarchy of measurable indicators, it became effortless to assert that "advanced" nations naturally led the world, while "backward" ones required guidance. What Buckle produced was not merely a history of civilisation, but a template, a data-driven vocabulary for judging cultures, a language of inferiority disguised as science, one that the empire would later deploy to reshape how Indians saw themselves.

5.5 BEHAVIOURAL DESIGN: INCENTIVES, SURVEILLANCE, CONTROL

While statistics and history were being marshalled to rank people and civilisations, philosophy was quietly providing the machinery to influence their behaviour. Jeremy Bentham reduced human action to a single calculus: individuals move toward pleasure and away from pain. If this was true, then behaviour was mechanistic, predictable, regulable, and open to design. Behaviour could be shaped through incentives and obedience could be maintained through the strategic distribution of pleasure and pain.

Bentham demonstrated that behaviour could be shaped not only through the direct application of incentives or pain, but through their anticipated possibility. Expectations, rather than force, governed conduct. His Panopticon, a building in which people could always be watched, induced behaviour as if they were being watched. Surveillance did not need to be continuous; it only needed to be conceivable. Authority could thus operate not through force, but through design.

Using such insights, Bentham's behavioural calculus shaped the design of British prisons, workhouses, and poorhouses, producing early architectures of surveillance and discipline. In this shift lay the intellectual seed of modern behaviour manipulation, the idea that conduct could be shaped by engineered expectations and more specifically by the architecture of possibility rather than direct coercion.

And thus, Britain acquired a righteous lens. People could be tabulated according to the attributes the British elite deemed important, measured against their own standards, and judged as "deviant" or "inferior" when they failed to match them. Once classified, they could be steered toward the "good" through behavioural conditioning, a moral project masquerading as science. The issue here was not that Britain had an ideal; it was that this ideal was not scientific but cultural, a projection of its own hierarchy, morality, and self-image, treated as universal law.

This behavioural logic did not remain confined to Britain. It travelled with its administrators and traders, carrying with it the belief that other societies, too, could be tabulated, judged, and corrected once their patterns had been rendered visible and measurable through the British ideal. By the time Macaulay arrived in India in 1834, the work of tabulation was nearly complete. For nearly seventy years, Company officials had been measuring and sorting India into neat administrative categories: literacy tables, caste schedules, land surveys, crime registers, and civilisational comparisons.

India had been rendered legible to British eyes, a civilisation allegedly without science, without useful knowledge, without industry, and without moral progress. These judgments were framed numerically against the British ideal, with low literacy, low productivity, high superstition, and high criminality. Whether these numbers accurately reflected reality mattered far less than the interpretive frame that British righteousness, and its misuse of statistics, made possible. Once India appeared "inferior" on Britain's new civilisational scale, the conclusion that it required British correction followed naturally.

5.6 MACAULAY AND THE ADMINISTRATIVE CAPTURE OF THE MIND

This is where Macaulay enters the picture. He did not create this righteous worldview; he inherited it. He viewed India from within that ideological vantage, a pre-fabricated frame in which British norms defined civilisation itself. His 1835 Minute on Education reads like the natural administrative solution to a diagnosis that this worldview had already produced.

In Macaulay's hands, the statistical portrait of India appeared as a civilisational deficiency in need of British correction. His proposals were built on the momentum of the governance tradition that had already shaped British civilisation, a tradition convinced that societies should be measured, ranked, and improved toward British ideals. When Macaulay called for an English-educated elite "Indian in blood and colour, but English in tastes, in opinions, in morals, and in intellect," he was not speaking from personal prejudice. He was speaking from within a civilisational ethos that saw itself as the apex of science-backed rational progress and regarded it as its duty to improve those deemed inferior. Macaulay was not the architect of this worldview; he was its instrument. The project belonged not to a man, but to a culture.

And what followed was not cultural exchange but cultural cannibalisation, justified by a century of pseudo-scientific comparisons that placed Britain at the peak of civilisation and India at the base.

The problem was not the use of statistics, but the elevation of British lived experience into a universal ideal against which all others were judged. A civilisation measuring the world in its own image could only ever find itself superior.

Thus Britain emerged as a society that embraced science yet fused righteous, egoistic overtones into itself, mistaking selective measurement for moral authority. It did not simply observe the world; it convinced itself that it had the power and the duty to remake it. If Hobbes opened the discussion, Petty supplied the arithmetic; if Buckle furnished the vocabulary of "advanced" and "backward" nations, then Galton supplied the mathematics of hereditary hierarchy; if Bentham provided the logic of inducing obedience, then Macaulay supplied the administrative blueprint.

5.7 ROMANTICISING EMPIRE

Not to be left behind, the poets and literary fashions of British society also romanticised the pedestal on which the British had placed themselves. Rudyard Kipling supplied the moral poetry. His famous claim that the Englishman bore the "white man's burden" to civilise the colonised was the emotional climax of the intellectual architecture Britain had spent a century constructing. Kipling translated Buckle's hierarchies, Galton's rankings, and Macaulay's prescriptions into a single, seductive moral. The empire did not dominate; it uplifted. Indians were not ruled; they were rescued. Kipling did for emotion what Macaulay did for administration. He naturalised inequality by framing domination as benevolence.

And thus, after the English Civil War shattered the divine right of kings, Britain rebuilt its political imagination around reason, measurement, and science. But this shift did not produce humility. Instead, Britain fused genuine scientific curiosity with a rising moral certainty that it stood at the apex of human progress and therefore had the authority, even the obligation, to reshape others in its own image.

By the nineteenth century, Britain believed it had unlocked the grammar of human behaviour, how to measure it, predict it, rank it, and steer it. Two centuries of refinement had honed this science on Britain's own population, embedding it into governance, policing, schooling, and social order. When the empire turned to India, the logic was already complete. It had counted, compared, and condemned; now it would correct. India would be judged against the British ideal and "improved" accordingly. What followed was a meticulous and quietly self-righteous project of cultural engineering.

5.8 MODERN MIRROR

As we saw, the British misused statistics to create social hierarchies, and on the basis of those hierarchies, they took decisions, shaped policy, and imposed their will. The danger did not lie in the numbers alone but in the assumptions smuggled into them. Statistics, subtle, deceptively simple, and easy to glorify, is uniquely vulnerable to human bias.

The empire believed its measurements revealed truth when, in fact, they revealed only its worldview. That temptation has not disappeared. Today, the same quiet distortions enter our surveys, algorithms, optimisation models, and dashboards. Bias seeps in not through malice but through presuppositions and worldviews we barely recognise, the reflexes of gender, class, language, and even colonial habit that silently shape how we collect, analyse, and interpret data.

5.9 GOVERNING BY METRICS

In today's data-rich world, governance increasingly operates through quantification. When population metrics themselves are imported, the underlying logic of governance is imported with them.

One of the first challenges in behavioural statistics is realising we do not actually know what we are measuring. Measuring physical quantities is straightforward. A ruler measures length, a scale measures weight, and a clock measures time. But the moment we turn to behaviour, emotion, trust, satisfaction, loyalty, happiness, or culture, the clarity evaporates. These are not objects; they are experiences. They do not exist in centimetres or kilograms. And yet modern statistics demands that we turn experiences into numbers. The act appears scientific, but it is deceptively simple, too simple. Without deep philosophical clarity about what exactly is being measured, we end up quantifying shadows, proxies, and convenient metrics instead.

Over time, subtly and unconsciously, the proxy becomes the goal. We start optimising customer satisfaction metrics such as the number of stars instead of actual customer satisfaction. Instead of establishing human connection, we focus on increasing likes and shares. Trust is reduced to retention scores. The map becomes the territory, and we confuse the metric being measured with the underlying phenomenon. The point becomes obvious in the extreme. True affection is not revealed by the number of heart emojis, and true love cannot be measured by the size of an engagement ring.

The British made the same mistake, intentionally or otherwise. They mismeasured India at every turn, defining literacy as English literacy, treating crime as whatever their own records captured, and counting productivity or education through standards that reflected Britain, not India.

5.10 WHEN MEASUREMENT IS OUTSOURCED

India, as a statistical civilisation, has yet to develop its own conceptual grammar for measuring behavioural life. The problem here is not technological infrastructure; it is epistemic.

Even today, the underlying ideas of what to measure and how to measure it remain imported. Our behavioural sciences rely on Western psychometrics, Western definitions of trust, satisfaction, or happiness, and Western survey logic that assumes people behave as they do in Europe or America. These frameworks travel poorly.

Today, India faces a deep problem. As the world becomes ever more data-oriented, governed by algorithms, optimisation models, and statistical dashboards, India increasingly measures itself with tools, standards, and techniques built elsewhere. India imports survey frameworks, behavioural metrics, economic forecasting models, psychometric scales, poverty lines, ESG templates, and even happiness indices designed in contexts that bear little resemblance to Indian reality. In doing so, it risks outsourcing not just computation but self-definition. When we measure ourselves through someone else's formulas, we inherit their assumptions, absorb their biases, and see ourselves in their light.

The danger is not only in importing Western logic but in optimising ourselves around it. Applied uncritically, these frameworks push us to optimise variables shaped by Western cultural assumptions rather than Indian lived reality. And once a variable is measured, it seldom stays passive; it becomes a tool of governance. The British measured India to rule it; today, institutions measure individuals, organisations, and communities to manage them. Performance dashboards run governments and schools. Risk scores guide policing. Credit scores decide who receives a loan. Health insurers track behaviour through wearable devices.

Consider the Western logic of happiness, which treats wellbeing as the product of individualism, the belief that a person is happiest when they feel fully in control of their life. In India, a civilisation built on dense social interdependence, joint families, shared obligations, and emotional cross-subsidy, this premise simply does not apply. And yet, when this metric is used, India is ranked predictably low on global happiness indices, a result so misaligned with lived experience that it comes across as comical.

But the real danger is not the ranking; it is the optimisation. Imagine India actually taking this metric seriously. We would begin "fixing" happiness by weakening family bonds, isolating individuals, reducing interdependence, and manufacturing Western-style hyper-individualism. In other words, to score well on a foreign index, India would have to engineer itself into a civilisation it has never been. Again, the absurdity becomes obvious in the extreme, but the point remains. Important metrics are not measuring India; they are measuring how well India imitates someone else.

The way forward, then, is for India to build the intellectual muscle to decide which experiences deserve measurement, how they should be quantified, and which require entirely new categories. It must abandon the habit of importing ready-made indices and optimising for someone else's vision of wellbeing. Even before measurements are made and numbers are generated, the concepts behind them must be meticulously interrogated. India must develop a measurement philosophy rooted in its own social logic, one that reflects interdependence, community, obligation, and collective flourishing. Only by thinking deliberately about what it measures and why can

India prevent optimisation from dragging it toward a cultural destiny not its own.

5.11 WHEN FOREIGN MODELS DEFINE RISK AND WORTH

In colonial times, the British treated the statistical averages of their own elite as moral centres and cast every deviation as defect. A similar logic governs the digital world today. Modern algorithms quietly turn "the average" into the desirable and "the deviation" into the dangerous. Recommendation engines push users toward statistically typical behaviour; hiring systems filter out profiles that diverge from historical patterns; predictive-policing tools flag outliers as potential risk.

First, the very idea of converging toward a single average is not inherently Indian. India is a civilisation built on plurality, linguistic, cultural, social, economic, and behavioural. Reducing this diversity to a single "representative" human flattens lived reality. Many Western statistical models assume a degree of population uniformity or behavioural homogeneity that simply does not hold in India. Applied blindly, they generate misleading patterns and flawed inferences. Such a gap offers India the opportunity to build a statistical grammar that treats plurality as foundational, developing models with multiple centres, multiple behavioural logics, and multiple truths. Such a foundation would not imitate Western centrality but emerge from Indian lived reality.

Second, even the "average" we now converge toward is not Indian. It is WIRED, shaped by Western datasets,

Western emotional cues, Western platform incentives, and Western behavioural assumptions about privacy, mobility, agency, and individuality. India still measures itself against Western standards, often without noticing that the profiling benchmarks themselves come from lives, histories, and psychologies very different from its own. What India needs instead is its own dataset to anchor its statistics. Not a single Indian average, but a constellation of averages that emerge from Indian plurality. This requires collecting behavioural data rooted in Indian life, defining its own conceptual categories, and benchmarking citizens against Indian realities rather than borrowed templates. Only then can the country move from imitating foreign norms to articulating its own.

Third, even when India uses fully local data, the model logic interpreting that data often inherits Western assumptions, embedded not in the data, but in the model families themselves. Many predictive frameworks were originally designed on Western behavioural patterns. For instance, in American datasets, dense neighbourhoods correlate with crime, so density becomes a risk flag. Applied in India, where density often signals community and safety, the same logic misclassifies neighbourhoods as "high risk," skewing credit scores, insurance pricing, and automated screening. The data may be Indian, but the risk logic is not.

To overcome this, India must build its own model families, predictive frameworks rooted in Indian behavioural realities rather than Western histories. This requires more than local data; it requires local logic. India needs algorithms trained on Indian social density, Indian mobility patterns, Indian informal networks, Indian credit behaviour, and Indian emotional cues. Only when the structure of the model reflects the structure of Indian life will the outputs reflect Indian truth. Without this, even Indian data will continue to produce Western conclusions.

To summarise, today India increasingly uses statistical models and predictive tools pioneered in the West. But in doing so, it risks losing something essential: the human realities that do not fit neatly into Western formulas. Statistics can be a powerful instrument for a pluralistic civilisation like ours, but only if the tools are shaped for Indian complexity rather than borrowed wholesale. At present, the West invents new metrics, new models, and new applications of these models, while India rushes to adopt them with minimal adaptation. The danger is silent but severe. We inherit their biases, absorb their assumptions, optimise for their categories, and end up evaluating ourselves against their ideals. Even basic measures like GDP distort Indian reality, ignoring the vastly different purchasing power, healthcare costs, family structures, and informal networks that shape Indian wellbeing.

To speak in numbers without being reduced by them, India must develop its own statistical grammar, one grounded in its own social logic, economic structures, and cultural truths. Otherwise, the country will forever remain on the wrong side of metrics designed elsewhere, metrics that were never built to capture India in the first place.

This is the underbelly of digital colonialism: statistics itself. This is the danger of an invisible discipline. It shapes the world without ever being seen. Formulas encode ideology, and ideology quietly becomes power. When we import Western algorithms, we inherit Western ideologies and Western destinies. But if we create our own, we speak to the world in our own frame. We measure ourselves against others on our terms, and we build the foundations from which a society draws its power, articulates its story, its standards, and its sense of what is just.

In an algorithmic era, to possess one's own ideology, one's own vision of the good life, the just society, and the flourishing citizen, India must possess its own algorithms.

6

From NEWSPAPER *to* SOCIAL MEDIA

6.1 CONTROLLING INFORMATION FLOW

The British Empire left India in 1947, but its gaze never truly did. Under that gaze, Indians had learned to see themselves through foreign eyes, to doubt their worth, to soften their voice, to seek approval from the very gaze that diminished them. Since independence, the buildings of governance were repainted, the railways renamed, the offices reoccupied. Yet even today, the reflex endures; we behave as though the gaze still lingers, performing deference long after the watcher has gone. The empire may have departed, but not the grammar of thought it left behind.

Seeking "English-speaking bride" and "English-speaking groom" still signals class. In courtrooms, the colonial echo of "My Lord" endures; in the bureaucracy, civil service remains the highest aspiration, draped in the old virtue of serving empire. Public-school manners and Western-educated polish still pass for competence, while regional accents are apologised for before being spoken. Even morality bears the same imprint: to be "God-fearing" remains praise, a colonial virtue, not native to India's spiritual lexicon.

We continue to think in the empire's logic and measure ourselves by the values it left behind. We use its vocabulary, aspire through its hierarchies, and conform to its "grammar of worth". We no longer need the coloniser's approval; we grant it to ourselves, through imitation. The empire no longer rules the land, but it still rules our minds.

These cultural reflexes are not harmless habits of speech; they are the sediments of the empire. As this chapter will show, they were carefully engineered into our collective consciousness. Thomas Macaulay had declared that material conquest was incomplete without mental submission. To command labour was one thing; to command imagination was another. The British sought not merely to extract India's resources, but to occupy its mind, to govern what a people could think, dream, or desire.

The true prize was not land or labour, but imagination itself, the inner landscape of thought. Their task was to police the invisible, to map, monitor, and finally master the Indian mind. To drain a civilisation of confidence until it saw its own reflection as inferior; to make its aspirations seem misplaced, its desires impure, its dreams childish. To rule not the body, but the sense of possibility and hope itself. That was the empire's most monstrous ambition.

6.2 DEFINING MEANING

The British understood that to control the thoughts of a civilisation, they first had to control its language. As the arbiters of English grammar and meaning, they could define what counted as proper speech, correct thought, and civilised expression. Language became their first frontier in the colonisation of Indian cognition. By setting the rules of grammar, they set the limits of imagination. For language does not merely describe reality; it defines the range of what can be thought about it. Once the empire decided what words meant and what logic counted as reason, it sanctioned not only the voice of the colonised but the structure of their thought itself, determining what could be said, what could be implied, and what could never be imagined at all.

So how does one, having seized language, curtail imagination? How does grammar become cognitive governance?

6.3 EMOTIONAL CODING

To this end, the empire fashioned one of its subtlest instruments, what I call emotional coding. Emotional coding was the deliberate engineering of feeling, the art of attaching select emotions to select words or sentences and rewiring the colonised mind's associations to induce predictable behaviour.

In psychology, such processes are understood as emotional or mental associations formed naturally through experience;

here, however, I use the term emotional coding to distinguish their deliberate and systematic infusion into Indian society as a tool of governance rather than an organic by-product of life.

The British engineered an information ecosystem in which emotional coding was foundational to advancing the colonial cause. Every message carried a preloaded sentiment: the coloniser was framed to evoke admiration and awe, while the colonised were conditioned to mirror inferiority and doubt. Pride was redirected upward, toward the empire; shame was turned inward, toward one's own identity and culture. From classroom lessons to newspapers, communication carried this embedded logic—superiority coded into the ruler, subservience into the ruled.

Words did not merely describe events; they coloured them. Riot, mutiny, native, and civilising mission were all encoded with hierarchy. The word rebel carried the presumption of criminality; the word native, the presumption of inferiority. To call someone rational was to make his reason sound English, calm, polished, and under control. To speak of mutiny was to imply betrayal, disloyalty, a moral failure disguised as rebellion. Reform meant obedience; law and order meant submission. The English language taught the colonised to see the world through the syntax of Western superiority. They learned to associate resistance with treachery and punishment with justice. The empire's lexicon became an invisible constitution, giving words reserved for the British a shade of superiority, and words for Indians a tint of inferiority.

Over time, these linguistic reflexes hardened into a learned moral order. Once words were framed, news stories began to prime expectation. Mutiny had been met with brutal punishment; the association endured. Each time the word appeared again, it summoned that memory, the warning that rebellion led to ruin. And so, through the colonial press and official reports, the very word "mutiny" was infused with fear and disorder, an emotion rehearsed until it became instinct. Readers no longer needed to be told what to feel; the emotion arrived with the word itself.

Through repetition, language trained reflexes and came to symbolise the desired outcome of the British ideological war. A British officer smiling over a dead tiger became a symbol of courage; a starving Indian family, an emblem of backwardness.

The British painted an emotional landscape that mirrored their hierarchy, progress and order reserved for themselves, barbarism and disorder projected onto the colonised. Through their teachings and newspapers, they did not merely transmit ideas; they transmitted emotion. They did not just control speech; they governed sensation. This was another of the empire's great innovations, to rule not by force but by feeling, to construct hierarchy through the architecture of emotion.

6.4 THE ARCHITECTURE OF EMOTION

The power of language was well understood and the British carefully guarded the Indian emotional landscape. A poem, a speech, or an idea charged with feeling, anger, awe, grief, or joy, travels faster and endures longer than the neutral or factual. History turns on this truth: revolutions, reforms, and empires have all moved on emotional energy more than on logic. The message that moves the heart always outruns the one that merely informs the mind.

To harness this power early, the British built the channels through which the emotional force of language would travel. They constructed the very networks through which information travelled, schools, presses, railways, and telegraphs, lacing them with emotional coding while silencing every channel they could not command. Each connection, from schoolrooms to churches, presses to post offices, railways to telegraph lines, became part of a circulatory system through which the empire's messages, morals, and moods, all encoded with emotion, could flow. It was not merely the infrastructure of communication, but the infrastructure of perception, ensuring that what was said in London could be felt in Lucknow.

Of course, it was not declared as such. This network was celebrated as one of the empire's "civilising gifts." Colonial rhetoric often grouped the postal service, the railways, the telegraph, and the English-language press as proof of imperial benevolence and modernity. The newspaper was hailed as the moral instrument of enlightenment and tools for rational debate. Indian thinkers saw the truth beneath the veneer. Dadabhai Naoroji called such gifts "mechanisms of extraction," R.C. Dutt likened them to "a vein connecting the body to its leech," and Gandhi, in Hind Swaraj, warned that the railways, press, and law had "enslaved the nation in the guise of service."

The empire's railways connected the body of India; its communication networks connected the mind. Both were hailed as gifts, and both were instruments of control. Emotional coding was the design; control of information networks was the machinery. One programmed feeling, the other distributed it. Together they formed the psychological factory of empire, a system that manufactured loyalty not through argument but through affect.

But language alone was not enough. For the English language to endure, it had to be adopted, and adopted by the Indian elite. The empire needed institutions that could lend it moral authority and social prestige.

To this end, missionary schools and colonial offices worked in tandem to sanctify English as moral speech. It was not only a tool of governance, a passport to advancement and a prerequisite for entry into power, but also a fashionable ideal, rewarded as a mark of civilisation and worn as a badge of honour. To speak English was to sound rational; to write it was to appear enlightened. Mastery of the language promised not merely employment, but elevation, a chance to rise above the "native" condition by imitating the ruler's tongue. The empire thus ensured that even when Indians spoke, especially those climbing the ranks, they spoke within its frame, using words that carried British meanings and implications.

And thus, the issuers of this language, missionaries, colonial officers, schoolmasters, editors, became the empire's emotional gatekeepers, shaping how entire populations interpreted the world. They decided which tones were respectable, which emotions were civilised, and which were vulgar or dangerous. Through this, the British managed the inner climate of the colonised mind, teaching subjects to associate calm with authority, restraint with reason, and agitation with disorder. To rule the voice was to rule the feeling, and through feeling, to rule belief itself.

6.5 GOVERNING THROUGH NARRATIVES

In time, the newspaper became its pulpit of reason: clipped tone, clean type, moral calm. Authority needed no drumbeat; it spoke in serif. The press now anchored English as the language of authority. Each headline, phrased in this moral cadence, reinforced the idea that truth itself spoke in English.

Readers began to associate reason with restraint, debate with dignity, and dissent with disorder. Over time, the newspaper became a moral compass, its calm cadence mistaken for truth itself. When colonial officers, magistrates, and professors all echoed the same editorials, the publication itself acquired sanctity. What the sahib read became what the city discussed, and soon, what the nation assumed. To be seen debating the day's paper was to appear modern and informed; to ignore it was to appear provincial.

Reading the press brought the colonised mind a step closer to the ruler's, creating the illusion of intellectual parity. The newspaper defined not only what could be said, but what could be safely thought. One did not have to agree with what was written; one merely had to be able to discuss and debate it.

The newspaper became the language of advancement: one had to read it, grasp it, and internalise its logic.

Imagine an Indian in such a conversation, debating, over a splendid lunch in the gardens, an editorial on the "primitive Indian mind." He argues in a language not his own, against a newspaper long since exposed as tactless propaganda, before an audience that listens not to understand but to affirm its superiority. Articulating dignity, he performs belonging while struggling to defend not only his civilisation but his own right to thought.

Thus, to be informed and able to articulate the thoughts the press prescribed was to appear civilised; to fall out of step was to appear ignorant. Over time, ambition did what coercion once did; Indians read to belong, and in belonging, absorbed subordination as progress. The empire no longer needed to enforce obedience; it wired it into pleasure and pain, into the difference between advancement and being left behind.

6.6 ENGINEERING NARRATIVES

Once narratives could be engineered to secure belief, control no longer required chains; it depended on circulation. The empire discovered that persuasion worked not only through what was said, but through when and how often it was said. Truth became a matter of rhythm. The information landscape, laced with emotional coding, was engineered around tempo, such that speed, repetition, and calibrated sentiment became tools of governance, not tools of truth.

The British realised that whoever spoke first fixed the frame; whoever followed merely argued within it, and always at a disadvantage. The telegraph became their nervous system, a wire that could outrun rebellion itself. A telegram from India in 1857 reported that Indian soldiers had attacked their British officers, a bare fact, stripped of context. By the time it reached London, it had acquired a name: the mutiny. In that single act of labelling, resistance became treachery, grievance became disobedience. From that moment, every alternative account was forced into defence, to deny or explain but never redefine.

The empire released its narrative early, a riot described as unrest; a protest framed as sedition. Even Bhagat Singh was portrayed in the colonial press as a political murderer, a man driven by passion, not principle. His own account, of course, was quite different.

The British narrative moved with precision. Its timing was as deliberate as its tone. Each report arrived before the facts could mature, fixing interpretation in advance. And the regularity of that rhythm did the rest. Each report rehearsed the same moral theatre: reason disciplining passion, law confronting disorder. Reading the same tone across multiple newspapers, the subject began to confuse consistency with truth. Familiarity became evidence; repetition became memory; memory became conviction.

To keep the machinery of emotional manipulation running, the empire did not need to invent new truths; it only had to rehearse old ones in new costumes. The same feelings were repackaged with fresh facts. What made this system durable was not novelty of argument but repetition of emotional content. Familiarity itself became persuasion. Psychologists now call this the illusion of truth: the more a claim is repeated, the more believable it feels, regardless of evidence. Each retelling of the empire's moral story, through reports, sermons, and editorials, reactivated the same emotional codes and linguistic frames discussed earlier. The elite repeated them in conversation, the educated echoed them in essays, and soon the colonised began to think within the same structure of belief. Through repetition, the empire transformed propaganda into common sense.

But persuasion did not end with rhythm; it perfected itself through anticipation. Each report arrived with its explanations ready-made, the British version of events appeared balanced, reasoned, almost weary with fairness. Every objection a reader might raise was already acknowledged and resolved. Dissent was not denied; it was domesticated. The empire's prose performed equilibrium: statistics softened emotion, moral appeals tempered power, sympathy appeared where dominance needed disguise. Even the letters to the editor played their part, polite disagreement printed under imperial supervision, proof that civility could stand in for freedom.

Thus, the reader was never forced; only convinced. He was made to feel that his agreement was the result of reflection, not influence. This was the genius of imperial narrative: it left no argument unanswered, no middle ground unoccupied. The British seldom asserted authority; they ruled instead through the comfort of plausibility and the illusion of debate.

6.7 THE ILLUSION OF DEBATE

Such a system of controlled communication had to be guarded, lest the illusion collapse. To preserve the illusion of debate, the empire built laws around language, ensuring that no voice could disrupt its rhythm. To further control circulation, the British empire also had to regulate who could print. The Vernacular Press Act of 1878 turned moral hierarchy into law. A single "seditious" paragraph could erase a newspaper; a sentence could summon prison. Indian-language presses such as Amrita Bazar Patrika and Kesari were harassed into silence, while English papers owned by loyal elites were treated as responsible voices. Freedom of speech became a licence that could be withdrawn; truth itself required permission.

If the intention had truly been open to debate, opposing voices would have been protected; but dissent was disorder, and control was needed to keep the English narrative neat and tidy. The entire ecosystem of discourse was performative, an elaborate architecture of order, posing as the voice of free speech and debate.

The Telegraph Act of 1885 carried the same insecurities into wire. Messages could be intercepted "in the public interest," a phrase so elastic it could stretch to contain any anxiety of power. The Post Office Act of 1898 extended this reach to private correspondence, allowing officials to open, read, and detain letters at will. Surveillance was moralised as safety. By the time of the 1931 Press (Emergency Powers) Act, censorship had become routine bureaucracy, editors forced to deposit securities, presses shut without trial, and films or pamphlets seized for disturbing the empire's "peace."

The Indian Press Act of 1910 tightened the rule further, branding dissent as undesirable and allowing presses to be closed without trial or justification. Freedom of speech under British rule was not a right but a favour, granted as a licence that could be withdrawn at any moment.

The empire conquered not only land but meaning, building railways to extract resources, and networks of information to impose belief. The British no longer needed chains or decrees; the Indian mind had become its own censor, the tongue its own gatekeeper. The empire's communication order, built on rhythm, repetition, and respectability, outlived the empire itself. By the time its laws faded, its logic had already settled into the Indian psyche.

The emotional residue of that vocabulary still runs deep in India's reflexes. To be English-speaking remains shorthand for being educated and respectable. Words like educated, modern, and civilised moralise imitation as progress, while local or vernacular still imply lack. Calm and English remains coded as reason; passionate and Indian, as unruly. The empire's emotional coding survives not in law but in language, a quiet grammar of deference that outlived the flag.

For when empires end, their grammars do not die; they migrate. The same logic that once governed print now governs platforms. Control no longer arrives as censorship but as curation, no longer through sedition laws but through design. What was once licensed speech has become ranked visibility;

what was once propaganda has become engagement. The empire of ink has become the empire of code, subtler, faster, and more intimate. The techniques that once shaped the colonial mind now shape the digital one: attention managed, emotion steered, visibility rationed, freedom performed. The telegraph has become the timeline, the algorithm, the new administrator of belief.

6.8 THE MODERN MIRROR

The empire's system of communication never vanished; it adapted. What the British once managed through presses, telegraphs, and law is now managed by social media platforms, feeds, and algorithms. Control has become more granular and distributed, no longer one message for the nation, but a customised stream for every individual. Each user receives a version of reality tuned to their preferences, fears, and impulses, while some voices are amplified and others filtered out. Selectivity remains, only now coded as moderation or design. The same rhythm of persuasion hums beneath our screens: where the press once taught Indians how to think, the feed now teaches them what to feel, each scroll a sermon without a preacher. Where the British licensed presses and editors, algorithms now license attention itself, deciding who is visible, who is amplified, and who is shadowed into silence.

Attention was always the true medium of control. In colonial India, it had to be cultivated; natives persuaded to read what the empire printed and to rehearse its opinions in clubs and drawing rooms. Today, it is optimised; tracked through clicks, seconds, and sentiment. Algorithms feed attention with emotion, shaping not only what people see but how they are meant to feel. Content is layered with affect, mockery to ridicule, seriousness to legitimise, pity to humanise, anger to mobilise, each tone calibrated to sustain engagement, not understanding. Every surge of activity carries an undertone of outrage, virtue, or fear, emotions selected not for truth but for traction. The empire once mastered emotional coding through prose; platforms now reproduce it through design.

The modern subject no longer waits for permission to speak; he competes for placement in the feed. The empire regulated voice through licences; platforms regulate it through relevance scores. Control has grown quieter but more total. Feeds promise dynamism, a living pulse of what the world is talking about, yet what becomes "hot" still depends on legacy media, corporate amplification, and state-approved narratives. The algorithm translates their authority into motion, ranking it as if it were popular will. What appears spontaneous is, in fact, curated volatility, a system that performs openness while reproducing hierarchy. Speed, repetition, and framing remain the architecture of narrative control.

Each serves a function: speed claims the moment, framing claims the meaning, and repetition makes both feel true. In the race to be first, accuracy has yielded to assertion. Platforms and financial media compete not to verify but to frame, to claim interpretive ownership before facts mature. The reward lies in timing, not truth. A report by a short-seller against a major Indian conglomerate triggered billions in losses, framing its claims to sensationalise rather than substantiate. The allegations remain contested, yet the market had already absorbed the verdict.

This is the new empire of velocity, where reputation and value can be rewritten in hours, and the first narrative becomes the final one. Investigations arrive as verdicts, headlines as moral judgments. And when they concern nations of the Global South, especially India, the reflex deepens. Praise feels rare; condemnation, habitual.

Attention is still directed towards stories that reaffirm Western virtue by rehearsing Eastern failure. The gaze has updated its technology but not its bias. Selective negativism has become the logic of contemporary Western media, when it comes to India, it amplifies the negative and mutes the positive. Such selective remembrance is another facet of the West's selective negativism, the same reflex that amplifies India's failures while editing out its contributions. The eye that once gazed upon the colony now decides what the world remembers. The pretence that once legitimised rule through infrastructure is now maintained through rule over imagination and recollection.

Language remains an instrument of hierarchy. The British once coded moral rank into words, civilised, native, reform. Platforms continue the work with subtler vocabulary: harmful content, trusted source, community guideline. The West keeps inventing new moral vocabularies to secure its interpretive power, post-truth, misinformation, toxic, unsafe, ethical AI, responsible corporate practices, inclusive innovation. These phrases sound virtuous but, like the civilising mission before them, position the speaker as guardian and the rest of the world as needing correction.

The same tint from colonial days endures; India is still seen through the grammar of lack. Slums, hunger, and chaos become its identity, while triumphs are rendered invisible beneath a sepia of poverty. This curated bleakness feeds the moral appetite of the world, from news to podcasts to film, a gaze that demands India remain its chosen site of suffering.

Like the empire, modern systems preserve the illusion of autonomy. The Raj discovered that visible control provoked rebellion, while managed freedom secured loyalty. Today, users can criticise power endlessly, so long as they do it within algorithmic boundaries. Visibility becomes the leash: reach throttled, comments buried, monetisation revoked, all without formal prohibition.

Dissent is no longer silenced; it is sorted. Voices that challenge authority are channelled into the "right" forums, subcultures, and online communities, spaces designed to absorb discontent rather than amplify it. The user vents, finds sympathy, cools down, and remains contained. Once opinions are neatly corralled, they can be categorised, extremist, reactionary, populist, anti-something, labels that neutralise critique without engaging it.

Control over communication has evolved, but its essence remains: to manage the emotional landscape of the populace. Yet every system built to curate thought eventually meets the one force it cannot predict: awareness. A recent BBC article asked why Indian youth appear less inclined to revolt than their peers in neighbouring countries. The answer may lie not in apathy, but in focus. Perhaps the silence toward anti-government protest highlighted in the article is not the absence of dissent, but the presence of purpose.

The architecture of control can manage imagination only as long as the mind remains captivated by the stream. Once attention turns away from the feed, the system loses its power. If that same energy is drawn toward creating and innovating, then imagination itself becomes resistance. By building what others prefer to dominate, India can shape the very future the West does not want to see materialise.

The task before India's youth is simple and radical: turn off the feed and dream the dreams others do not want them to dream.

7

From HUMILIATION *to* SUBORDINATION

The previous chapters traced how empire learned to govern not only through force, but through systems of measuring populations, shaping narratives, and controlling the terms through which reality itself was interpreted. Once behaviour could be anticipated and meaning could be framed, power no longer needed to rely on force. The next step was to direct and mould the subject itself. Control became far more efficient when it operated internally, eroding confidence rather than provoking resistance.

The ideal subject was one stripped of self-belief, uncertain and dependent. Revolt is costly and unpredictable, whereas self-doubt is stable and enduring. When worthlessness takes root, obedience becomes common sense.

It was at this point that humiliation emerged as a preferred emotional anchor through which rule was stabilised. Anger or fear provokes resistance, but shame encourages compliance. To govern millions with minimal force, dignity itself had to be corroded. Subjugation worked best when it felt natural, internal, and deserved.

Colonial power therefore ruled not only through guns, but through glances. A bowed head, a delayed reply, and a humble tone became instruments of empire. Material expropriation of land, taxes, and labour was only the surface. Beneath it ran a colder plan: to make a nation unusable to itself. Humiliation was not a side effect of empire; it was one of its most effective technologies—one whose effects have not faded with time.

7.1 THE RITUAL OF SUBMISSION

Under British rule Indians were treated not as citizens to be trusted with rights but inferiors to be civilized, corrected and governed. The colonial project did not stop at extracting wealth; it extracted dignity. It replaced a culture rich in self-governing confidence and spiritual identity with a trained habit of submission. Law, court, and school alike taught one lesson: Indians were to be ruled, not represented. Every petition and seal became a rehearsal of inferiority.

The only way to approach authority was through petitions and "considerations," rituals of submission disguised as procedure. Asking for one's rights had to be performed as a favour. To speak too directly was insolence. Every plea required an introduction, a title, and a tone of gratitude. Even the request for fairness had to be made gracefully and, preferably, in the Queen's English. Courtesy became the precondition for justice; submission, the price of being heard.

The wound ran deeper than economics. It was a loss of moral standing, of being seen as less than human in one's own land. Colonial governance turned humiliation into an instrument of order, leaving people stripped of legal standing to raise grievance. The Raj made humiliation a feature of governance.

The emotional toll was lasting: shame, dependency, fear of arbitrary power, the absence of equal standing as citizens. It created subjects who waited for approval instead of citizens who expected justice. As the next chapters will show, the psychology of permission outlived the empire itself.

7.2 THE MORAL CAMOUFLAGE OF THEFT

Empire required not only profit, but moral legitimacy. To plunder and still believe itself virtuous, Britain needed a moral alibi. It found one in humiliation. As Shashi Tharoor noted in An Era of Darkness, the British did not humiliate India merely to rule it; they humiliated to justify what they stole. Material plunder demanded an explanation that felt noble. The victim had to appear broken so that the thief could seem benevolent.

Colonial ideology became a form of moral accounting: extraction recast as benevolence, conquest as civilisation, exploitation as duty. The full weight of Britain's guilt was redirected outward, pressed into humiliating those it ruled. To keep its conscience intact, the empire needed India to appear degraded — so that every act of theft could be told as uplift.

They had to tell their children, and their children's children, that it was not cruelty but duty; not greed but guidance. That they were civilising, not stealing. It was the story that allowed the coloniser to sleep at night.

This was the double movement of imperial psychology: outwardly, humiliation disciplined the ruled; inwardly, it absolved the rulers. Every improvement proclaimed by the Raj, a permit granted, a relief announced, a favour bestowed, served both purposes: to refine the subject into obedience and to redeem the ruler from guilt.

A civilisation that wished to see itself as moral first had to invent inferiors to remain so. The British mastered the engineering of humiliation as precisely as they engineered their machines—turning shame itself into a technology of rule. Once they learned that guilt could be hidden inside virtue, humiliation became not an act but a system. It was institutionalised through schools, courts, and the press—each designed to reproduce deference as habit.

That architecture of control endures. Today, its instruments are digital rather than clerical: algorithms that steer attention, shape opinion, and calibrate emotion with mechanical precision. The obsession to ensure that the subject remains predictable—mentally guided, emotionally compliant—continues the same imperial impulse in subtler form.

7.3 MACAULAY AND THE INSTITUTIONALISATION OF HUMILIATION

Once the empire had discovered that moral superiority could conceal material guilt, it began to institutionalise humiliation. Thomas Babington Macaulay's Minute on Indian Education turned that moral theatre into policy—a blueprint for behavioural reprogramming written in the language of those who believed civilisation could be gifted. His call to create "a class of persons Indian in blood and colour, but English in tastes, opinions, morals, and intellect" transformed loyalty into curriculum.

Schools would not merely teach arithmetic: they would teach inferiority as the price of advancement. Curriculum, examination, and comportment codified deference: English diction, polite posture, measured tone. Indian identity itself was recast as something to be corrected—a degraded raw material to be refined into the British ideal. What was once pride became something to be polished away.

7.4 HOW HUMILIATION WAS DELIVERED

The previous chapters showed how empire learned to measure behaviour and shape narrative, making control quieter and more precise. Once communication and meaning in India were brought under British control, the imagination followed. When the space of imagination was colonised, humiliation emerged as the most efficient form of subjugation.

As this humiliation was introduced into the Indian psyche through a psychological pattern that closely resembles what we now recognise as behavioural reorientation. The surrounding environment, including institutions, education, administration, and everyday social interaction, was shaped in ways that diminished confidence and reinforced subordination.

These practices were repeated across daily life and gradually normalised. Over time, humiliation ceased to feel exceptional and became routine. Submission no longer felt imposed from outside; it began to feel appropriate. The subject learned not only how to obey, but how to interpret obedience as reasonable and even necessary behaviour.

These environmental and behavioural change techniques were integrated into a coherent system in which identity, behaviour, and aspiration were engineered together. At that point, empire no longer needed to humiliate continuously. It had learned how to reproduce itself.

7.5 THE MODERN MIRROR

The empire's methods never vanished; they migrated. What once licensed who could print now decides who is visible, who is amplified, who disappears beneath the scroll. The flag became a social media platform; conquest became curation. The new colonies are cognitive, not territorial. Just as the nineteenth-century empire used schools and newspapers to set moral agendas, today's platforms use algorithms and advertising to define what is "acceptable," "inclusive," or "responsible." Power still defines virtue, and virtue still justifies power.

7.6 THE PERSISTENCE OF SUBORDINATION

The Indian is still caricatured in global culture with the same lazy shorthand: the exaggerated accent, the head wobble, the curry smell. A billion lives compressed into a punchline. The Indian man, once portrayed as the obsequious "Babu," the barbaric "aboriginal," the exotic half-naked yogi, or the snake charmer, now returns in global meme culture as the "cringe" archetype: awkward, over-eager, socially inept. The colonial logic survives: too servile to command respect, too primitive to deserve it. The tone has shifted from superiority to comedy, but the condescension remains, the same old hierarchy, now delivered with a smile.

The Indian woman, meanwhile, is caught in another trap: she is painted as either too modest to be modern, mocked for her conservatism, restraint, and "primitive" beliefs, or too modern to be respectable, condemned by her own society for overstepping invisible lines, as if modernity itself required severing her familial and cultural roots. Both are distortions of the same gaze, a gaze that mocks like a bully to hide its own unease. From a comfortable distance, the West turns an ancient civilisation adapting to the modern age into a mockery, a reflex inherited from its colonial past.

The mockery performs no description; it performs reassurance. It sustains a hierarchy in which the West remains the arbiter of taste, and India, regardless of achievement, remains the punchline. The coloniser's tone survives in laughter for the Indian man and pity for the Indian woman. What was once ridicule in print is now ridicule in pixels, the same gaze, only faster.

History, however, has its own little ironies. What the empire once coded as inferiority — a need to learn, to be civilised — India later turned into fluency. What brings some comfort in India's history is that the very instruments that disciplined India also prepared it for success. The habits of petition, gratitude-driven bureaucracy, procedure, examination, and documentation that once served as chains later became ladders, producing a people fluent in paperwork, argument, and endurance. The empire's instruments of control became the subcontinent's instruments of ascent.

The colony learned the grammar of power so perfectly that, when the age of globalisation arrived, it could outperform its tutor. The empire wanted clerks; it produced CEOs. It wanted interpreters; it produced innovators. Thomas Babington Macaulay's class of intermediaries became the global managerial class — fluent in the protocols of empire, competent in its logic, and indispensable to its machinery. China may have matched India in population and technical skill, but India inherited the empire's true engine of advantage — English. It was never the language of freedom, but it became the language of power on which the British built systems of subordination. Indians mastered its use and, consequently, the mechanisms of bureaucracy itself.

The same demographic mocked online leads global business and technology; Indian households abroad rank among the world's highest by income. Indian women, have quietly rewritten the script — leading corporations, building startups, and shaping the frontiers of everything from artificial intelligence to medicine, from Silicon Valley to Singapore. Their success is not a stroke of luck or circumstance but a reflection of the social fabric they carry, families that invest in education, communities that prioritise perseverance, and a culture that, despite its diversity, still holds together. Yet none of this alters the tone of the caricature.

Even such success could not straighten the posture the empire had bent, for the world still freely speaks down as if the lesson of hierarchy had never ended. The ritual of humiliation has not vanished from international life; it has merely been repackaged as diplomacy. Former colonisers and their allies now perform virtue through lectures, on human rights, environment, gender, and governance, delivered to the very nations once denied dignity. India, even as it feeds, vaccinates, and powers hundreds of millions, is still addressed as a pupil in the world's moral classroom. Western governments issue advisories, panels publish reports, and journalists pose questions in a tone once reserved for viceroys — condescension dressed as concern.

The muscle of moral camouflage remains strong, the colonial instinct to recast misplaced moral superiority as self-anointed duty, even after leaving the Global South ravaged and famished. The stage has changed from the durbar to the UN, but the choreography endures, the same raised eyebrow, the same assumption of moral entitlement, echoing the condescension of the empire. In the age of algorithms, the colonial lecture has become a press release, a stage performance, a social media post.

A striking example of this lingering impulse lies in the narrative of women's safety. A subject that should have been unremarkable in an ideal world yet remains one of the most urgent and uncomfortable realities of ours. It has become one of the West's favourite mirrors for painting India in shades of danger and disorder. The narrative is repeated so often that it has hardened into cliché. Yet the data tell a more complex story. On a per-capita basis, India's reported crime rates against women are lower than in many Western nations from which these judgments originate. The question, however, is not whether India fares better or worse, but why attention itself is so selectively directed.

Flawed statistics, inflated by absolute numbers instead of proportion, are magnified into moral lessons about India, while equally troubling patterns in Europe and North America are quietly recast as "social challenges." The result is not reportage but ritual, the same tendencies of superiority seen in the colonial era in which humiliation becomes the instrument of reassurance for the other. Stories with the same context are "complex" or "contextual" when they occur in the West but become "systemic" or "cultural" when they occur in India.

7.7 NARRATIVE ASYMMETRY

This is narrative asymmetry: the systematic amplification of flaw and failure, paired with the minimisation or erasure of dignity and achievement. It is humiliation sustained by narrative design, the modern extension of the empire's moral theatre.

Narrative asymmetry operates not only in how India is criticised, but in what the world remembers. Even positive remembrance, when it comes, is filtered, partial, and conditional. In World War II — a conflict often retold with obsessive fidelity to detail — one in every three soldiers who fought for the British Empire was Indian. Yet in most "accurate" Western depictions of that war, especially those made for mass audiences, Indians are invisible. As historian Yasmin Khan observes, this silence represents "the erasure of the colonial soldier from Britain's memory of war." Santanu Das echoes this, noting how the empire's subjects "fought, suffered, and died in a war that later forgot their faces."

The absence is not accidental; it reveals how memory itself can be colonised. When accuracy demands acknowledgment, but ideology demands amnesia, history chooses convenience over truth. Their sacrifice vanishes from the cinematic record, as though their courage were inconvenient to remember. The contributions that would dignify India are forgotten; the portrayals that diminish it are amplified. The positive is erased in silence, while the negative is performed in stereotype.

Why are Indians deleted from stories that claim to be complete? Perhaps because to show them would mean conceding that the empire's survival, and much of Europe's freedom, rested on those it ruled. To admit that the colonised defended the coloniser would dissolve the moral supremacy on which the imperial narrative depends. It would shatter the empire's moral architecture, for it would demand a vocabulary the coloniser never permitted: the brave Indian soldier who fought and died for the democracy and freedom of the British. That such erasure persists even today, in films celebrated for their accuracy, suggests that the imperial narrative has not ended, only adapted.

Even as the world evolves, the West, given control of technology—the global media architecture in this case—seems unable to relinquish its oldest craving: the urge to use its technological advantage to feel virtuous by rendering others inferior. It seems that once it has tasted the high of false moral superiority, it cannot stop returning for another dose. What began as moral justification, the excuse that once legitimised extortion and rule, has become the dependence itself: a civilisation addicted to feeling superior at the cost of others.

This same addiction to righteousness, which the empire once fed through distance, is now indulged by platforms through intimacy. The command no longer shouts; it suggests. The modern subject is not coerced but curated, shaped by gentle nudges, guided by invisible recommendations. The next three chapters trace this new dominion: how communication, culture, and dissent were re-coded for digital obedience.

What was once the press is now the platform. What was once missionary education is now content virality and moderation. Censorship has become cancellation. The vocabulary has changed, but the psychology endures: to shape what is visible, to prescribe what is admirable, and to punish what is inconvenient.

Humiliation, once a tool of governance, has become the grammar of discourse. It tells billions how to speak, what to laugh at, and when to apologise. The empire that once extracted wealth now extracts attention, but the deeper extraction continues: the quiet, continuous theft of dignity and identity, rebranded as civility itself, the smile that conceals subordination.

8

From CULTURAL PROGRAMMING *to* ALGORITHMIC SOCIALISATION

In the previous chapters, we traced how the British arrived as traders, then became monopolists, and finally evolved into rulers. When trade alone did not deliver the profits they sought, they turned to a more reliable source of extraction: the Indian way of life itself. The more dependent Indians became on British systems and approval, the more stable and lucrative the colony became. Once culture was redesigned, rent could be extracted across every sphere of life, embedding the colonial economy into the everyday experience of Indian society.

This chapter examines how Indian culture was systematically redesigned to enable that extraction.

8.1 CULTURE AS A RESOURCE TO SHAPE AND EXTRACT

The British first reshaped the Indian emotional landscape. They seized the channels of communication, ensuring that the stories Indians told about themselves, and the stories the world told about India, flowed through a British filter. Prestige, civility, rationality, and progress were all recast in British terms. Humiliation became an instrument of control, eroding confidence and inducing doubt.

With emotional authority weakened and communicative authority captured, the British could now internalise humiliation and subjection. To achieve this, they reshaped the Indian environment so that conduct could be managed and identity redefined. The British had at their disposal an entire infrastructure of power: institutions, bureaucracy, law, newspapers, and schools. Through these instruments they rearranged the conditions in which Indians lived and acted.

Once that environment was engineered, the ideal colonial subject emerged almost automatically: one whose behaviour could be governed and whose identity could be shaped, a subject who would comply under the mere gaze of authority rather than resist.

By reshaping culture, the British could govern not only the conduct of their colonies but also their consumption, aspirations, and economic life. When a civilisation's sense of livelihood and dignity is defined from the outside, every path to progress passes through the coloniser, allowing rent to be extracted at every stage and locking the colonised into a self-reinforcing cycle of dependency and impoverishment that subsidised the comfort of the imperial centre.

In an age where we live inside curated digital worlds, revisiting how culture is shaped is vital, lest we allow environmental design to determine our destiny.

To understand how Indian identity was replaced with the ideal British-Indian, we will retrace these methods through the lens of contemporary behavioural science and the psychology of behavioural redesign. The British application of these principles bears an uncanny resemblance to the models of behaviour change we recognise today. In modern contexts, such models are presented as tools for improvement—to break harmful habits, deepen learning, or encourage positive outcomes. But the science itself is neutral; it becomes benign or destructive only in the hands of the wielder. In India, however, this science was used not to elevate but to subordinate. It became the basis of a comprehensive programme of cultural transformation—one that aligns with modern theories of behavioural change with a precision that is difficult to ignore.

They began by classifying and categorising Indians through British epistemologies and definitions. Diagnosis became devaluation, where Indian identities were diminished and recast as deficient. Re-orientation followed, presenting British values and conduct as the ideal to strive for. Reinforcement was built into the machinery of institutions, incentives, punishments, and law. Finally came internalisation, when the hierarchy became self-governing, self-policing, and taken as the natural order of things.

8.2 THE ARCHITECTURE OF CULTURAL REENGINEERING

Much of this period in Indian history has been written about from different facets: administration from one angle, race science from another, education reforms from yet another. What has rarely been done is to see these actions as components of a single behavioural system. Rather than offer an exhaustive catalogue of every episode of cultural imposition, this chapter follows the deeper behavioural architecture of control through the five steps of behavioural redesign. The aim is to understand how cultural engineering was operationalised and how its underlying logic now runs through the digital architecture of Indian life. The first step in that architecture was classification.

Step One: Classification

Every act of domination begins by defining what is to be dominated. Before the British could govern India, they first had to make it legible. The census became the first instrument of control, not an exercise of cultural curiosity but a prerequisite for power.

A civilisation that was fluid, diverse, and self-organising was recast into categories that could be named, counted, ranked, and managed. Complexity was not reduced because it could be understood, but because it had to be administered. Here, the epistemic issue raised earlier becomes decisive. Those who define the categories define reality. The British exercised this power in two moves: they chose the dimensions along which India would be measured—caste, race, occupation, criminality—and then imposed their own definitions and standards upon them. A scholar was considered "educated" only if he bore British credentials; mastery of Indian sciences or literature was dismissed as superstition.

Measurement was never neutral. The British counted only what their worldview recognised, and ignored what they did not comprehend or found inconvenient. The numbers they produced were laced with bias, prejudice, and self-righteousness. Once quantified, these categories hardened into official truth.

The absurdity of both the intention and the manner in which this enterprise was carried out is best seen in Herbert Risley's anthropometry, his belief that caste could be ranked by the width of a nose. A few millimetres of cartilage became a scientific justification for caste hierarchy, aimed at aligning Indians to a British ideal of appearance. The measurements have long since been discarded, but the hierarchy endured.

What made this especially consequential was the British belief in the heritability of traits. Under the logic of Galtonian heredity and the rise of racial science, behaviour and status were recast as inheritable. A census entry could condemn an entire lineage. Caste became hereditary: fixed, documented, and transmitted through administrative authority. The rigidity of caste, as we recognise it today, was not merely the legacy of an ancient institution but a colonial construction.

Thus, what we treat today as "backward" or "tribal" was often the product of British administrative invention. E.A. Hutton and others extended Risley's logic by re-classifying entire communities as "tribes" and treating ancient, respectable groups as inherently suspect. These labels did not arise from any intrinsic social reality, but from the British assumption that the only legitimate citizen was the one who was permanently settled—and therefore permanently taxable.

India's social landscape contained diverse ways of life—mobile pastoralists who moved with their herds, agro-forest communities who combined farming with forest resources, shifting cultivators who rotated land to preserve fertility, and itinerant groups who travelled seasonally. These ways of life had sustained their communities for generations.

The British did not attempt to understand these traditions, whether out of administrative convenience or economic ambition, and reinterpreted longstanding ways of life as deviance. Movement became "refusal to settle." Foraging became "inclination to theft." Seasonal migration became "lawlessness." A thousand distinct modes of living, too complex for the colonial imagination and inconvenient for revenue extraction, were flattened into crude stereotypes: "primitive," "unproductive," "dangerous," or "criminal by instinct."

The Criminal Tribes Acts took this logic to its extreme. Entire communities were designated "hereditary criminals," not on evidence but on the colonial belief that behaviour was biologically inherited. Once a group entered the census under this category, the stigma became permanent. Law shifted from investigating offences to policing identities. Camps, movement passes, fingerprinting, and surveillance became routine instruments of administration for these communities.

It is important to emphasise that these categories did not emerge from within India but were imported and imposed. India was made legible through foreign frames, and legibility through foreign definitions is always the first step toward submission. A society forced to answer in another's language eventually begins to think in that language. Those who define the categories define reality.

The census was never meant to accurately describe India; it was meant to divide it. And once divided, each fragment could be devalued on its own.

Step Two: Devaluation

Once identity is seen and classified, its worth can be devalued. The act of naming is followed by the act of judging. If the British had stopped at categorising India, their rule would have remained a mapping exercise. But they went further. Each category was assigned a verdict. Every labelled group acquired a stereotype that defined its social value.

Martial races were emotional, undisciplined, and fit only for soldiering under British command. The so-called educated Indian was merely schooled in superstition, rote, and dogma rather than knowledge. Indian religion was irrational, philosophy impractical, and traditions disorderly.

Indian commerce was corrupt and irrational. Its rulers were short-sighted and unfit for governance. Artisans and craftsmen were obsolete, their skills dismissed as relics of a dying past. Indian society was primitive and backward. Nothing met the British ideal of civilisation.

Each group was diminished in relation to a British standard—judged selectively on the one attribute that guaranteed inferiority. Individually, each stereotype felt plausible. Collectively, these judgements produced a single conclusion: that the entire civilisation lacked intrinsic worth.

Further, once identity was classified and judged, it could be divided. Classification allowed the British to govern a civilisation through division and fragments. As groups were separated into castes, tribes, "martial races," criminal communities, and backward classes, they could be administered, taxed, and policed separately. No collective Indian identity remained that could resist. Potential rebellion dissolved into dozens of smaller conflicts. Law and revenue were enforced not on individuals but on categories. Governance became effortless. The state no longer needed to understand Indian society; it only needed to manage the labels it had created.

Separate electorates turned identity into political division. Voting rolls were separated by religion and community; citizens could vote only for candidates of their own category. The electoral process itself became an instrument of separation. National unity became impossible by design because identity had become the organising principle of power. In each case, whole communities were governed not by their actions but by their classification. Policing, recruitment, and representation were reorganised around labels.

The Indian identity, once classified, judged, and divided, could now be ranked. The Martial Races doctrine turned identity into military hierarchy. Some communities were declared naturally brave; others were dismissed as inherently weak or cowardly. Recruitment into the army was deliberately ethnic and caste-based.

The same logic extended into economic life. New intermediaries of land and revenue—zamindars, taluqdars, jagirdars—were empowered from selected communities. Some castes and tribes were granted land and revenue rights; others were deliberately excluded. Patronage flowed along identity lines: certain communities received trading and shipping contracts, railway and opium concessions, and access to commerce and banking. The civil service and legal professions were dominated by a narrow educated class considered "collaborative" and "English-minded." This was not a coincidence but a systematic, documented design of governance. Classification turned into an inherited political and economic order.

But the deeper success of classification was psychological and political. Communities began to defend the identities and ranks assigned to them. The struggle shifted from resisting colonial rule to rising within the colonial hierarchy. Divide-and-rule now had an epistemic engine. Differential treatment kept groups competing rather than uniting. The

British appeared less as conquerors and more as referees. Each group looked to the empire for approval, resources, and status, always in relation to other categories. The hierarchy no longer needed to be actively enforced; it began to sustain itself. Classification allowed the empire to govern fragments instead of a civilisation, each fragment invested in the very categories that kept them apart.

A culture loses its authority when its interpretation of itself is replaced by an external, devaluing judgement. This was the key psychological pivot of cultural domination. It was not prejudice but an organised technique of power. Modern psychology has names for these strategies—status degradation, delegitimisation, epistemic injustice. In simpler terms: cultural gaslighting. A system tells you that your lived truth is invalid, that your history is unreliable, that your knowledge is defective. Once this doubt enters the mind of a civilisation, it becomes fertile ground for external authority. If your traditions are the problem, the coloniser becomes the solution. If your way of life is defective, imitation becomes the aspiration.

Once the British weakened the Indian frame of meaning, they offered a new identity to step into. This was the third step of cultural engineering, and it is the most misunderstood. Many historians treat British reforms as administrative modernisation or civilising mission. But reform is the wrong word. What happened was cultural realignment. Having dismantled the Indian scale of value, Britain installed itself as the reference point of civilisation. Indian identity was not merely judged; it was redirected.

Step Three: Reorientation

A society that has been convinced of its inferiority becomes vulnerable to guidance. A subject whose past is devalued will naturally wonder what they should aspire to be. The British then filled the vacuum they had created. They supplied the very ideals against which Indians were expected to measure themselves.

English became the prestige language. Rationality was redefined as that which followed British methods. Civility and gentility—Victorian inventions, not universal virtues—became the standard of refinement. Speech, manners, self-presentation, and even emotional expression were judged through British norms. Schools became instruments of cultural discipline and trained students in manners, posture, and accent. Newspapers shaped public narrative and public life. The law reshaped how Indians argued, reasoned, and proved their claims. The army reoriented value from courage toward obedience; the civil service rewarded conformity over initiative. Every institution of the Raj became an apprenticeship in British behaviour.

For the emerging aspirational classes, this reorientation produced a profound psychological shift. Ambition became tied to imitation. Advancement required fluency in British manners and expectations. Petition culture taught the performance of deference. Missionary and government schools standardised self-presentation, manners, and endurance without protest. Victorian morality spread as the acceptable code of conduct. A route away from the devalued native self now existed, but only through the British system.

The incentives rewarded imitation, permission-seeking, and dependence rather than originality, confrontation, and autonomy. The goal was to learn one's place within the system and behave correctly according to its standards. t is important to note that these ideals toward which Indians were reoriented were never universal. They were not even the ideals that governed British society itself. In Britain, courage, innovation, and dissent were admired; entrepreneurs were encouraged to take risks; the military rewarded audacity; and politics embraced confrontation.

A country that had once defied and resisted its own monarchy was teaching its colonies that obedience and restraint were virtues. The very qualities that fuelled British progress—ambition, experiment, rebellion—were discouraged in India. Risk-taking that built British industry was condemned as disorder in the colony. Industrial innovation was praised in Britain but restricted and tightly controlled in India. Protest and press freedom were increasingly protected as constitutional norms at home but criminalised as sedition in the colony. In England, mysticism, imagination, and romanticism were celebrated as high culture; in India they were pathologised as irrational. Even the rule of law bifurcated into two systems—one for the rulers and one for the ruled.

The claim that the British "exported their culture" to India as a sign of civilisation is a profound misrepresentation. They did not export British culture; they exported their expectations of subjects. The virtues Britain valued for itself were denied to the people it ruled. Colonial behavioural norms were not a gift or an exchange of ideas. They were a method of ensuring that no Indian could become too independent, too ambitious, or too close to the ruler.

If classification and devaluation weakened the cultural anchor, reorientation provided the new compass. The next step was simpler: engineer an environment where behavioural alignment was incentivised and any deviation from that aspiration came with a cost.

Step Four: Reinforcement

Once the new identities had taken hold and the older ones had been weakened, the British turned to the next stage: reinforcement. This meant engineering the environment so that compliance became the most sensible choice. Reinforcement was the architecture of predictable behaviour. It designed the incentives of daily life: who would be rewarded, who would be admitted, who would be heard. The empire structured the environment like a maze, steering the colonised to act in ways that benefited the coloniser.

Reinforcement operated along four intertwined channels: institutional, social, economic, and psychological. Schools and courts drilled the posture of inferiority (institutional). Prestige and respectability were reserved for those who conformed (social). Jobs, land, and contracts flowed through the obedient (economic). Over time, the craving for approval and fear of exclusion wired colonial obedience into emotion itself (psychological).

Every system the British built—schools, courts, bureaucracy, communication—was designed to channel behaviour toward alignment with empire. Upward mobility flowed only to those who walked the path the British engineered. The environment itself did the teaching. Schools rehearsed discipline. Courts rehearsed deference. Newspapers rehearsed civility. Offices rehearsed hierarchy. English fluency became the prerequisite for employment. The press defined respectability. The law demanded a posture of submission. In such an environment, obedience no longer felt imposed but became the path of least resistance.

Institutions acted not as administrative bodies but as behavioural workshops. Curricula, procedures, forms, and protocols rewarded the posture of inferiority and penalised deviation through exclusion—from jobs, contracts, office, prestige, and upward mobility. These institutions were not built to serve a population; they were built to shape one.

As advancement depended on behaving correctly, a new social hierarchy took shape: those closest to the British ideal were closest to success. Rewards for subservience were not simply symbolic but were material too. Jobs, land, contracts, and office were tied to obedient and loyal behaviour. The empire channelled material success as a teacher of habit. An English-speaking clerk rose quicker than the most learned scholar. A polite petitioner climbed faster than a principled dissenter. Even resistance was graded. Those who opposed empire in the format the British found acceptable were confined in palatial prisons; those who rejected that format and acted in unpredictable ways were sent to Kala Pani. Economic incentives codified the behaviour the state wanted.

Colonial rule demanded predictability and had no use for spontaneity. It needed a population that could be forecast like an equation. Through law, ritual, and humiliation, the state cultivated subjects who anticipated the wishes of authority before authority spoke. This was the real triumph of governance: a mind that performed obedience before being asked.

Reinforcement succeeds when behaviour becomes routine. A society repeatedly exposed to the same incentives imitates automatically. It learns the safest posture and the most profitable habit. Once a generation had been trained to see success through the British lens, the next inherited the behaviour without remembering the punishment that produced it or the culture it displaced. This is the moment when colonial behaviour begins to reproduce itself. This is the moment when colonial thinking becomes reflex.

The craving for approval, the fear of exclusion, the desire for prestige—all were engineered into the emotional reward system of society. Behavioural psychology confirms the empire's intuition: behaviour changes belief faster than belief changes behaviour.

Once behaviour becomes habitual, a deeper transformation begins. The system no longer needed enforcement; it had begun to police itself from within.

Step Five: Internalisation

Internalisation is the final stage of cultural engineering—silent, invisible, and self-perpetuating. It is the moment when the coloniser no longer needs to enforce his norms because those norms have migrated into the minds of the colonised.

Once enough Indians adopted the colonial mindset, something remarkable happened: they began to enforce the norms themselves, often with more zeal than the British. Schools became sites of cultural reproduction, punishing deviations that would previously have gone unnoticed. Bureaucrats became stricter than their colonial supervisors. Elites judged one another by Western standards to signal optimised alignment. Communities competed with one another within colonial ranks and hierarchies.

The logic of imitation began to accelerate through network effects. When a neighbour succeeded within the colonial structure, imitation became rational. When a relative gained status through English education, the family redirected itself toward Western schooling. When submission to bureaucratic authority became the price of advancement, communities reorganised their ambitions. People abandoned identity for stability and culture for opportunity. The British no longer had to enforce their norms once society began to enforce them from within. Culture itself became the gatekeeper.

Internalisation is the most advanced form of subjugation. A civilisation that once drew confidence from its own systems now seeks validation from the very power that judged, ranked and shaped it. You do not need soldiers to discipline someone who polices himself. You do not need censors when the imagination remains within the boundaries you created. You do not need repression when your approval is their highest aspiration. You do not need chains for someone who has forgotten he was ever free. The most complete conquest is the one in which the conquered forget they were ever conquered.

8.3 THE MODERN MIRROR

The colonial logic of cultural programming never truly ended. It has reappeared in digital environments that capture and engage attention, condition behaviour, and administer identity with algorithmic precision.

Borders and armies are no longer required; platforms now govern identity, behaviour, attention, and desire. The modern empire does not need administrators—it has algorithms, nudges, metrics, feeds, and prediction models. Behavioural science is woven directly into the architecture of technology. Scholars like Zuboff, O'Neil, and Harari have already traced how digital life shapes cultural identity, so there is little need to repeat their analyses here.

The only point relevant for us is this: the colonial logic remains fully intact for India. Indians are still classified—this time as users, consumers, segments, and risk scores. They are still judged—ranked by influence metrics, algorithmic preference, and proximity to Westernised digital norms and customer-lifetime values. They are still reoriented—through influencers, advertising streams, curated news cycles, and aspirational narratives. They are still reinforced—through likes, shares, amplification, algorithmic reward, or deplatforming. And they still internalise—through content engineered for emotional precision, timed dopamine hits, behavioural triggers, and manufactured outrage or desire.

The system may seem new, but the logic is not. This is algorithmic socialisation: digital platforms have become the latest infrastructure through which behaviour is shaped, identity is modulated, and aspiration is redirected. As we shall see in the next chapter, it is through this cultural alignment that markets are structured and value is continuously extracted from Indian society.

Digital platforms have become the latest infrastructure through which Indian behaviour is shaped, Indian identity is modulated, and Indian aspiration is redirected. Indians, still carrying colonial reflexes, adopt external systems without doubt, hesitation, or fear. China builds its own digital infrastructure because it remembers the cost of dependence. Russia builds because dependence is existential. India however comfortably outsources its digital foundations because it was trained to see dependence not as a threat, but as normal. The psychological logic of colonial behaviour remains: we expect culture to be shaped from outside. And thus, we trust foreign platforms and their content without pausing to question the cultural cost. The instinct is still to align, adopt, and conform—and in doing so, we continue to risk letting our imagination be shaped by external powers. A civilisation that loses agency over its own imagination does not merely risk exploitation; it risks stagnation and the quiet erosion of self-belief.

India knows this pattern well. The famines, extractive taxation, engineered dependencies, and slow corrosion of confidence—internalising inferiority—were not historical accidents but structural outcomes of an earlier era of cultural reliance. Yet the reflex to depend endures. Comfort in dependency remains the most persistent legacy of colonial internalisation.

In the digital age, colonialism will not defeat India through armies or imposed rule, but through cultural inertia, through habits that outsource thinking, innovation, and vision. This raises an uncomfortable question: can India withstand another wave of colonialism—this time algorithmic, optimised, and invisibly embedded in the architecture of daily life?

If not confronted, the result is predictable. Others will shape what Indians buy, how they spend, what they watch, how they speak, and eventually how they govern themselves. The Indian citizen risks being reduced once again—not as a subject of the crown, but as a loyal and satisfied consumer of the platform.

9

FROM COLONIAL SUBJECTS TO SATISFIED CONSUMERS

The final stage of colonialism was the monetisation of desire. As seen in previous chapters, Indian identity and desire was shaped, disciplined, and normalised through narrative control, cultural engineering, and the steady elevation of British ideals as universal benchmarks. Once Indians had internalised inferiority and their culture had been reoriented to climb British aspirations, the empire settled into a comfortable and durable form of rent collection through consumerism. The British had to simply exploit and monetise this desire. Rent, or lagaan, was collected through desire embedded in everyday choices and habits.

This desire was supplied and stabilised through British branded goods and services. British products came to confer legitimacy and status, operating as visible markers of refinement and modernity, while indigenous culture and way of life increasingly signified lag. What Indians aspired to wear, consume, and emulate was defined abroad—constantly renewed, deliberately scarce, and priced beyond necessity. Status had to be repeatedly purchased, allowing rent to be collected indefinitely.

Clothing offered the clearest illustration. Indigenous dress was gradually displaced by British suits; greater status attached to suits made from British cloth; still greater prestige followed garments tailored in London itself. Each step up the ladder signalled proximity to British standards, while deepening economic and cultural dependence.

This process locked Indians into a downward spiral. Indigenous ways of life were steadily devalued, while British lifestyles were elevated as markers of refinement and progress. Participation required constant alignment with foreign standards—of dress, taste, and behaviour—each priced above necessity. Those unable to keep pace were marginalised or excluded from social and fashionable circles. With every shift in fashion, legitimacy had to be repurchased and aspiration renewed, allowing rent extraction to repeat itself continuously.

Pierre Bourdieu observed that taste functions as a social code—misrecognised as personal preference but operating as an instrument for maintaining hierarchy. In colonial India, this logic was embedded directly into consumption. The cost of British fashions lay not in substance but in symbolism. There was nothing intrinsic in British cloth that justified its inflated prices; the material itself was often Indian, with only the branding rendered British. What was being purchased was recognition. Status, legitimacy, and social acceptance adhered to the label rather than the object. Each act of consumption therefore carried a built-in loss: distinction originated elsewhere and expired quickly. Consumption became a perpetual effort to keep pace with externally defined standards, ensuring that rent could be extracted repeatedly and without end.

Every mechanism of colonisation explored earlier—monopoly, behavioural engineering, humiliation, control of communication, cultural redesign, and dependency—ultimately converged into a single outcome: a civilisation positioned to finance another civilisation's future through consumption. India was reorganised to channel its wealth, attention, and aspiration outward, largely toward the West. Consumption became the primary measure of advancement, while the capital generated by that consumption funded innovation, prestige, and future power elsewhere. Participation in British-defined markets, brands, and lifestyles functioned as a steady transfer of resources and imagination.

Advancement was experienced socially and symbolically within India, yet it generated little foundational capacity of its own, reinforcing a cycle of dependency in which the ability to build indigenous technology—or to author an independent future—remained structurally constrained. Beyond perpetual rent extraction and status subordination through ever-changing fashions defined abroad, the consequences of consumerism and adoption were deeper and more enduring. First, they produced an illusion of advancement. Second, they steadily extinguished indigenous capacity for innovation.

9.1 THE ILLUSION OF ADVANCEMENT

By the late colonial period, India appeared increasingly "modern" by global standards. Railways crisscrossed the subcontinent, English education expanded, bureaucratic systems functioned with efficiency, and British institutions governed commerce, law, and administration. To those participating in these systems, the experience felt progressive. Life appeared ordered, connected, and aligned with what was presented as the most advanced civilisation of the time. This visibility of imported systems created a powerful sense of advancement. This appearance of advancement was reinforced daily through consumption—of goods, habits, and lifestyles that signalled proximity to imperial standards.

Yet this appearance concealed a deeper asymmetry. A society could operate sophisticated systems without exercising authority over them. Indians learned to work within British institutions, to follow British procedures, and to excel according to British standards. Mastery of these systems signalled competence, intelligence, and upward mobility. But the systems themselves—their design, priorities, and ultimate objectives—remained external. Participation did not translate into control.

This gap between experience and authority produced an illusion of progress. Advancement was felt through access and inclusion, yet structural power remained unchanged. Railways enabled movement, but not ownership. Courts delivered order, but not sovereignty. Education produced clerks, administrators, and professionals fluent in British frameworks, but rarely authors of independent ones. Modernity was experienced at the level of function rather than foundation.

Over time, being modern came to mean appearing modern. Progress was measured by visible adoption rather than by the ability to build or own modern systems. Visible interfaces—institutions, infrastructure, social practices—appeared advanced, while the structures governing them remained firmly imperial. India looked modern, but its ability to define economic, industrial, or technological direction remained constrained.

Adoption gradually reshaped how progress was understood. Fluency in British systems came to be seen as achievement. Aligning with British systems replaced independent thought as the primary pathway to success. Social mobility depended on proximity to imperial norms rather than the creation of indigenous alternatives. To adopt efficiently was considered pragmatic; to attempt independent systems appeared unnecessary or unrealistic.

Cultural validation reinforced this orientation. British education, manners, and lifestyles became markers of refinement and civilisation. Participation in imperial institutions conferred recognition and legitimacy, even when it offered little real authority. A class emerged that felt advanced, educated, and modern, while remaining structurally subordinate within the imperial order. Confidence grew from inclusion rather than control.

Exclusion made the limits of this arrangement unmistakable. Spaces that symbolised refinement and progress—clubs, hotels, and institutions—were frequently closed to Indians. Signs declaring "Dogs and Indians Not Allowed" were not anomalies but expressions of a governing logic. Indians could adopt British habits and dress, yet were repeatedly reminded of the boundaries of their acceptance. Modernity was permitted as imitation, not as equality.

As seen in earlier chapters, this arrangement was not accidental. Colonial infrastructure was designed to integrate India into imperial circuits while retaining strategic control elsewhere. Railways facilitated administration and extraction. Legal systems stabilised governance. Educational institutions produced intermediaries capable of operating the empire efficiently. These systems created the experience of advancement while preserving asymmetry of power. Infrastructure, though functional, was never neutral. Authority followed those who defined its purpose.

The result was a form of advancement that was visible, experiential, and socially rewarded, yet fundamentally limited. India appeared to be progressing while remaining dependent on external authorship. Competence expanded without sovereignty. Presence increased without power. The colony looked advanced. That was not a failure of colonialism. It was its design.

9.2 THE PERMISSION TO INNOVATE

The other consequence of consumerism and the habitual adoption of foreign fashions, systems, and technologies was the erosion of India's ability to innovate and to author its own technologies—and, ultimately, its own future. Colonial India was not lacking in ingenuity, effort, or enterprise. What it lacked was permission. Permission to create, to imagine, to think independently, and to explore paths not already sanctioned. The imperial order did not envisage India as a site of creativity, but as a territory to be governed, supplied, and consumed.

Consumerism presupposed dependence. Indians were trained to adopt British technology, goods and services, while the authority to define ones remained external. Gradually, innovation in India came to mean adoption and refinement rather than the pursuit of foundational or breakthrough creation.

Colonial rule actively discouraged experimentation. Risk-taking outside sanctioned domains was rarely rewarded and often rendered illegitimate. Capital, legitimacy, and institutional support flowed toward activities that complemented imperial priorities — administration, extraction, trade, and efficiency — rather than toward ventures that might challenge control over economic direction or technological architecture. Innovation that threatened to alter underlying structures was systematically marginalised. What flourished instead was activity that improved performance within limits already set.

This produced a peculiar condition: visible industriousness without directional control. Enterprises could grow, markets could expand, and productivity could increase, yet the fundamental terms of progress remained externally authored. Innovation existed, but it was bounded. Active, but not sovereign.

The pattern was visible across the colonial economy. Indians were encouraged to innovate in trade, logistics, finance, and intermediary services—domains that facilitated imperial commerce and administration. Merchants, brokers, and agency houses prospered by connecting local markets to imperial networks. In textiles, Indians participated as traders, processors, and suppliers of raw materials, while mechanised production and global pricing power remained concentrated elsewhere. Plantations allowed for managerial and operational innovation, but ownership, market access, and strategic control lay beyond indigenous hands.

Infrastructure followed the same logic. Indians worked as clerks, engineers, and contractors within railway systems that were technologically sophisticated and administratively complex. Local enterprise improved scheduling, maintenance, and efficiency. Yet the railways themselves were not instruments of indigenous industrial integration. Their routes, financing, and priorities were designed to serve imperial extraction and control, not to foster autonomous production or innovation ecosystems.

Professional domains reflected similar boundaries. Law, administration, and education expanded rapidly, producing a class of skilled Indian professionals fluent in British systems. Legal practice, journalism, and publishing flourished within permissible limits. These fields rewarded interpretation, narration, and administration, not authorship of foundational frameworks. Enterprise was real, competence visible, and social mobility achievable—but always within inherited structures.

Across these domains, innovation was encouraged where it stabilised and extended the existing order. Efficiency was rewarded; independence was not. Activity was welcomed when it complemented imperial systems, not when it sought to redefine them. The boundary was rarely announced, but consistently enforced through capital flows, prestige, and institutional validation.

Over time, this conditioning reshaped capability itself. A society trained to operate within externally defined systems develops proficiency in execution rather than conception. Innovation narrows to optimisation. Origination comes to appear unnecessary, risky, or implausible. The question shifts from what might be built to how well what already exists can be implemented.

This was not a sudden collapse, but a gradual erosion of the ability to innovate. Innovation was redirected, constrained, and contained. Indigenous capacity withered quietly, as repeated execution replaced experimentation and alignment replaced ambition. What remained was a civilisation capable of intense activity, yet increasingly unfamiliar with the act of defining its own future.

This asymmetry becomes clearer when viewed alongside developments in the imperial core. In Britain, these same domains—industry, infrastructure, law, media, and education—were sites of origination. Legal doctrines were debated and codified, not merely applied. Industrial technologies were invented, standardised, and exported. Media institutions shaped narratives rather than operated within inherited frames. Railways were not only administered but designed as instruments of national industrial integration. Innovation in the imperial centre defined direction; innovation in the colony improved execution. The difference was in the ability to exercise authorship.

9.3 THE RISE OF CEREMONIAL POWER

When authorship is denied, innovation shifts away from origination toward permitted forms of execution and refinement. The boundaries of what was permitted were set by British-sanctioned institutions, norms, and systems. What emerged was a structure of ceremonial authority—highly visible, socially recognised, and carefully rewarded, yet strategically hollow.

Colonial rule reorganised them the Indian elite. Leadership was permitted, but only of a particular kind. Authority was extended to those who mirrored imperial norms, internalised imperial values, and operated comfortably within imperial frameworks. Western education, English fluency, institutional affiliation, and cultural alignment became prerequisites for recognition. Those who adopted these markers were elevated as leaders, administrators, and representatives of progress.

This elevation carried status, prestige, and material reward. Titles were conferred. Institutions were staffed. Public roles were granted. Palaces, courts, clubs, and universities became sites where power was displayed and acknowledged. To the public eye, authority appeared intact. Indian figures occupied prominent positions. Governance seemed locally embodied.

However, this authority was largely performative. Decision-making power remained external. Indian elites were permitted to exercise administration, representation, and symbolism, not authorship in technology, economics, or law. They governed within rules they did not write, implemented policies they did not design, and presided over institutions whose ultimate purpose lay elsewhere.

Crucially, leadership itself was filtered. Only those sufficiently trained in Western modes of thought were considered fit to lead. Indigenous systems of knowledge, governance, and political imagination were steadily displaced, not by force alone, but by disqualification. Leadership that did not conform to imperial standards was rendered unintelligible, unprofessional, or dangerous. In this way, colonial power narrowed the field of possible leaders long before narrowing the field of possible policies.

This had lasting consequences. When authority is granted selectively, the nature of innovation changes. Aspirants learn not to challenge the system, but to resemble it. Success becomes a matter of alignment rather than originality.

Leadership stops being an act of vision and becomes a performance of competence within inherited norms. The horizon of imagination contracts accordingly.

Ceremonial power thus served a stabilising function. It created the appearance of participation without the substance of control. It rewarded loyalty without granting sovereignty. It ensured that those who rose within the system had already internalised its limits. The empire no longer needed to suppress alternative visions aggressively as such divergent and independent thinking rarely reached positions of visibility in the first place.

What emerged was a class that appeared powerful, confident, and modern—yet whose authority was recognised socially rather than exercised strategically. They occupied offices without owning futures, administered systems without defining them, and embodied progress without directing it. Power was no longer something one wielded; it was something one performed. This transformation did not weaken colonial rule. It refined it. By converting authority into ceremony, empire ensured continuity without constant intervention.

The symbols of leadership remained Indian. The substance of power did not.

9.4 PUPPET KINGS OF THE LATE EMPIRE

The systematic denial of autonomy and agency created a persistent tension within the colonial order, particularly among Indian elites. Political authority had been hollowed out, strategic control withdrawn, and the capacity to shape long-term direction removed. To stabilise this demand for agency, the British offset the loss of real control by prioritising display and grandeur. Princely states, nawabs, landlords, and selected indigenous elites were encouraged to convert constrained authority into spectacle.

This took the form of lavish residences, ceremonial courts, titles, and lifestyles that signalled prominence without conferring control. Grandeur substituted for agency. Visibility replaced sovereignty. As substantive power narrowed, it was increasingly expressed through display—palaces instead of policy, ceremony instead of strategy, consumption instead of authorship.

Even technology itself became part of this economy of spectacle. British technologies—automobiles, electrical systems, communication devices, and industrial machinery — circulated among Indian elites less as instruments of independent capability than as symbols of modernity and alignment. Possession signalled relevance and proximity to imperial standards, not control over production, standards, or technological direction. These technologies were consumed, displayed, and admired, but rarely mastered as systems that could be replicated, modified, or owned end-to-end. Technology functioned as ornament rather than capability — visible proof of modernity without the means to reproduce it independently.

This visibility mattered. It redirected ambition. Aspiration flowed toward status, recognition, and proximity to imperial culture rather than toward authorship or independence. These figures became objects of admiration and envy, reinforcing the belief that success consisted in rising within the hierarchy rather than challenging its structure. The spectacle of elite prosperity helped normalise the order that sustained it.

Yet this elevation remained tightly bounded. Lavishness did not translate into autonomy. These elites did not control their own destinies; they remained vassals, removable and replaceable at imperial discretion. Succession, territorial authority, revenue rights, and even personal security depended on continued compliance.

Crucially, this lack of autonomy extended beyond politics into the economic and technological domain. These elites did not control the direction of industrial development, technological capability, or economic architecture. Even where wealth was substantial, it did not translate into the freedom to define systems, set standards, or invest independently in foundational innovation. Intermediaries could administer, distribute, and consume, but not originate.

In this arrangement, governance became a matter of management. Indigenous elites implemented policies they did not design, operated institutions they did not own, and upheld systems whose ultimate purpose lay elsewhere. Their role was to maintain order, collect revenue, and embody legitimacy—not to shape futures.

The result was a paradoxical form of power: highly visible, socially reinforced, and materially rewarded, yet strategically hollow. These figures were kings in ceremony and lifestyle, but vassals in authority. Their prominence did not signal autonomy; it signalled incorporation. Their success demonstrated the rewards of alignment, not the possibilities of independence.

By elevating intermediaries while denying them control, the colonial system transformed leadership into administration and ambition into compliance. Power no longer required constant imposition. It was mediated, localised, and normalised through figures who appeared sovereign while functioning as custodians of an order they did not command.

9.5 SEEING THROUGH THE SPECTACLE

Yet these displays of grandeur did not deceive everyone.

A small number of Indians who reached the upper tiers of colonial society began to recognise the hollowness of spectacle without agency. Having experienced rank, access, and visibility, they discovered the limits of status that could be withdrawn, overridden, or ignored at will.

Some responded by attempting to author institutions rather than merely inhabit them. Jamshedji Tata's insistence on building a hotel that did not exclude Indians marked a refusal to accept British standards as the gatekeepers of dignity and modernity. The Taj Mahal Palace asserted that modern institutions, standards, and dignity could be created locally rather than accessed through British permission. It was an act of institutional authorship in a domain where Indians were expected to consume, not define.

Similar instincts appeared in heavier and more contested domains. Indigenous steel production, most notably through Tata Iron and Steel, sought control over the material foundations of modern power rather than participation in imperial trade alone. The effort faced sustained skepticism and obstruction from colonial authorities and British industrial interests, revealing how uncomfortable indigenous authorship made the imperial system.

In shipping, the Scindia Steam Navigation Company represented a direct challenge to British logistical dominance. It was founded explicitly to break imperial monopoly over maritime transport, a sector critical to trade, pricing, and movement. The company faced discrimination in insurance, port access, and regulatory treatment, operating under conditions that made scale difficult and sustainability precarious. Its struggle exposed how indigenous innovation was structurally constrained and deliberately contained.

Banking followed a similar pattern. Institutions such as Punjab National Bank and Bank of Baroda emerged to provide Indian-controlled credit and reduce dependence on British exchange banks. They supported merchants and local enterprise, yet remained confined to narrow commercial roles. Control over monetary policy, industrial finance, and large-scale capital flows remained firmly external. Financial autonomy was partial, localised, and deliberately capped.

Even within emerging sciences, patriotic figures such as Homi J. Bhabha emerged at the edge of empire. Formed within imperial institutions and having reached their highest levels of recognition, they were nevertheless unwilling to mistake prestige for power. Their response was not deeper alignment, but authorship. Through institution-building, long-term vision, and insistence on indigenous control over foundational capabilities, their initiatives laid the groundwork for what would later become India's nuclear and strategic technological capacity.

These efforts were neither accidental nor naïve. They reflected an understanding—rare under colonial conditions—that real progress required ownership of assets, systems, and direction. Yet they remained isolated, expensive, and vulnerable. They did not displace British monopolies, nor did they evolve into a self-reinforcing ecosystem of indigenous innovation.

Their importance lies in what they revealed. Those who reached the apex of colonial hierarchy often discovered it to be empty. Visibility could be purchased. Prestige could be displayed. But agency had to be self-generated.

The colonial order had trained both the public and the elite away from this realisation. Through consumerism, spectacle, rank, and reward, ambition was redirected toward appearance rather than authorship. Success was framed as access to British systems, not independence from them. The few who attempted ownership instead of imitation stood out precisely because the system had been designed to make such aspirations appear unnecessary, impractical, or dangerous. These exceptions do not soften the argument. They sharpen it. They show that even when Indians attempted to reclaim authorship, they did so against a structure that preferred them as consumers, intermediaries, and symbols—never as architects of their own future.

9.6 A CIVILISATION TRAINED TO ADOPT

Over time, these arrangements did more than organise markets or institutions. They shaped behaviour. A civilisation was trained—quietly, persistently—to adopt. Adoption came to be understood as intelligence. Alignment with British standards brought rank, recognition, and legitimacy, marking such choices as safer and socially rewarded. Imported systems were perceived as proven and reliable. To follow British norms was prudent; to imitate them sensible; to adopt what already existed abroad appeared rational, respectable, and fashionable.

Originality carried uncertainty and risk. It demanded capital, protection, patience, and permission—conditions that were rarely available. Attempts to build independently invited scrutiny, resistance, or quiet obstruction. Over time, ambition adjusted itself accordingly. The instinct moved away from invention toward adoption, from authorship toward recitation.

As these habits repeated across generations, they hardened into instinct. Technology adoption became normal. Dependence on foreign goods and services came to appear obvious, aspirational, and natural. Borrowing became the norm. Originality was forgotten.

9.7 THE FINAL COLONIAL FORM

And so, colonialism reached its final form. A stage in which the desires of the colonised were continuously satisfied by the coloniser. Control no longer depended on occupation or overt enforcement. It rested on alignment—quiet, habitual, and mutually reinforcing. What people came to want aligned seamlessly with what the imperial system was structured to supply.

At this point, Indian consumption itself fuelled the empire. As wealth flowed outward, Britain financed new technologies, industries, and fashions. Indians participated willingly as aspirational consumers, experiencing the sensation of progress without acquiring its substance. The loop sustained itself. Desire generated consumption; consumption generated rent; rent financed further innovation elsewhere.

India had been reorganised into a civilisation that continuously funded another civilisation's future—an enduring system of rent collection sustained not by force, but by consent.

9.8 MODERN MIRROR

While the British departed and India gained political independence, the underlying structure of rent collection did not disappear. It adapted. In the digital era, the same logic re-emerges through technological dependence, with rent extraction intensifying as everyday life becomes embedded in platforms, standards, and systems designed elsewhere.

The contemporary form no longer requires the language or ceremony of empire. It operates through markets, platforms, and profit. Control is exercised not through administration, but through ownership of technology stacks, standards, and interfaces. Colonial administration has been replaced by platform governance; imperial trade by digital ecosystems; lagaan by recurring digital rent. The form has changed but the underlying mechanisms and logic remain intact.

In colonial India, the subject was gradually remade into a consumer. The Western gaze continues to view India primarily as a market, and Indians as consumers within it. Under British rule, desire was trained through fashion, etiquette, and consumption cycles defined abroad. This training was carried through the communication technologies of the time—newspapers, advertising, publishing, and public discourse—largely controlled by British interests. In modern India, the same mechanism operates through Western technology.

Technology today is presented as progress itself. Consumer electronics, software platforms, and digital services arrive framed as markers of growth, modernity, and inclusion. Possession signals relevance. Adoption signals advancement. Yet these technologies do more than provide tools and convenience. They also function as channels through which narratives, aesthetics, and aspirations are transmitted. The same platforms that enable communication and work also deliver films, images, lifestyles, and ideals that shape taste, behaviour, and cultural reference points.

Through these systems, Indians are exposed continuously to Western modes of living, dressing, consuming, and desiring. Fashion, entertainment, and lifestyle circulate through the same digital infrastructure that carries messages, media, and labour. Western brands are normalised as aspirational, while legitimacy increasingly depends on visibility within platforms governed elsewhere. Consumption follows naturally. What is seen, admired, and rewarded is what is purchased.

Consumer electronics, platforms, and services therefore function not merely as tools, but as markers of modernity. Annual upgrade cycles replicate the colonial rhythm of aspiration and expiry. Devices remain technically functional, yet rapidly lose social value. Obsolescence is cultural before it is technical. Legitimacy decays on schedule, and status must be repurchased.

Digital platforms now perform the cultural work once carried out by imperial fashion, etiquette, and the colonial press. They rank visibility, reward alignment, and quietly withdraw attention from those who fall behind. Algorithms determine what circulates, what trends, and what disappears. Status is continuously measured, distributed, and revoked. As before, legitimacy expires quickly and must be reacquired, ensuring continuous consumption.

Progress is experienced through acquisition. Modernity is signalled through possession. To remain current is to stay aligned with externally defined cycles of innovation, fashion, and obsolescence. India participates enthusiastically in these cycles, even as their direction, standards, and cultural logic are authored elsewhere.

The result is a form of advancement that is visible, lived, and socially rewarded, yet structurally constrained. Like before, India appears technologically dynamic while remaining dependent on systems it does not design, govern, or control. Technical competence expands, but sovereignty does not. Participation deepens, while authorship remains external. The form has changed, but the underlying logic has not.

As in the colonial economy, India occupies the position of a market rather than a partner. It is indispensable to global firms not as a site of origination, but as a site of scale. Users, data, attention, and revenue flow into platforms owned and governed elsewhere. Participation is deep. Ownership is shallow. India consumes systems it does not define, operates within standards it does not set, and finances innovation whose trajectory it does not control. The asymmetry is structural.

Colonial lagaan has been replaced by digital rent. Subscriptions, licensing fees, cloud usage charges, platform commissions, advertising revenues, and data extraction ensure continuous outward flows of value. Ownership has been replaced by access. Payment is recurring. Exit is costly.

Indian consumption now finances research, development, and experimentation abroad. New technologies are imagined, built, and refined elsewhere, then sold back as products, services, and platforms. Progress feels real locally, yet it remains imported, conditional, and revocable.

In the colonial era, indigenous authorship was deeply discouraged and systematically constrained. Enterprises that sought control over foundational capacity encountered regulatory bias, capital denial, and exclusion from critical infrastructure. Independent authorship was rendered costly, fragile, and failure-prone.

The modern digital economy reproduces this constraint in subtler form. Foundational technologies are opened to Indian participation largely through partnership—as vendors, implementers, or distributors—while ownership, deep know-how, and architectural control remain tightly guarded. Barriers to entry are raised through standards, intellectual property, and platform control, making independent authorship not impossible, but prohibitively difficult. As before, adoption is encouraged while autonomy is discouraged by design.

Hence, as before, adoption is framed as intelligence. Imported platforms arrive pre-legitimised, perceived as reliable, safe, and inevitable. To adopt them appears rational; to build alternatives appears risky. The burden of proof lies not with the foreign system, but with the local attempt to replace it. Alignment is framed as pragmatism. Borrowing as efficiency. Originality once again carries uncertainty without institutional protection.

This conditioning extends to talent and leadership. Just as colonial education produced clerks fluent in imperial administration, modern education produces engineers, managers, and analysts fluent in global technology stacks. Indians excel at operating complex systems they do not own. Many rise to senior roles within global firms. The visibility is real. The success is real. But the nature of power remains managerial rather than generative. Foundational decisions—architecture, standards, and strategic direction—are typically set before they arrive.

This visibility also performs a stabilising function. The elevation of Indian executives within Western technology firms creates the appearance of representation and inclusion, even as control over core systems remains unchanged. These figures occupy prominent positions and are celebrated domestically as symbols of success. Their presence reassures audiences that authority has diversified and that power has travelled upward. Yet their role is structurally intermediary rather than sovereign: they administer, optimise, and scale systems whose underlying logic was defined elsewhere.

In this sense, visibility substitutes for control. Leadership becomes ceremonial rather than generative. These executives mediate between Western platforms and Indian markets, translating priorities downward and legitimacy upward. Their prominence normalises dependence by giving it a local face. As in the late colonial period, authority appears Indian while authorship remains external. The structure does not rely on deception by individuals, but on a system that rewards representation without transferring power.

Within India, innovation is energetic but bounded. It flourishes at the application layer—apps, services, aggregators, integrations—while control over foundational layers remains external. Indian firms build on platforms they do not own, within ecosystems they do not govern. Each layer above pays rent to the layers below.

Even when domestic platforms scale, they are largely encouraged and sustained because they connect Indian users, merchants, and enterprises more efficiently into global technology grids—cloud infrastructure, operating systems, payment rails, advertising networks, and data pipelines—whose ownership and governance lie elsewhere.

This is why global technology firms are eager to "support innovation" in India. They are not funding challenges to foundational control; they are expanding demand for their own platforms. Innovation is encouraged precisely where it deepens dependency. Capital flows readily into startups that extend Western ecosystems, but rarely into firms attempting to redefine them. Indigenous companies that seek deeper authorship remain exceptions rather than the rule—and are forced to grow without the same institutional backing, narrative legitimacy, or capital density.

The contrast becomes clearer when viewed internationally. China, for all its constraints and imperfections, treated technological sovereignty as non-negotiable. It built parallel digital ecosystems—its own search engines, marketplaces, social platforms, cloud infrastructure, and payment systems—ensuring that usage translated into domestic capability.

When foundational AI models emerged in the West, China responded by authoring its own. These models were not perfect, but they were indigenous in control, governance, and trajectory. China adapts and replicates external ideas, but it does not accept permanent dependence through import and adoption.

India, by contrast, largely adopts. It integrates readily, often without contest or counterproposal. This absence of even attempted authorship reflects the internalised legacy of colonialism: a system trained to treat adoption as intelligence and origination as unnecessary risk.

The result is a familiar pattern. Innovation exists, but it is filtered. Creativity is abundant, but channelled. Entrepreneurship thrives within boundaries set elsewhere. India appears technologically dynamic while remaining structurally dependent, keeping pace with technology even as the technology itself remains an imported dependency.

What persists beneath all this activity is an unspoken assumption of subordination. India remains unable to design and manufacture advanced chips, to independently build and govern large-scale data-centre infrastructure, or to author digital platforms that fully shape and serve its own social and commercial ecosystems. The foundational layers of the digital stack remain external.

Across the complete technology chain—from the silicon chip to computation devices such as phones and computers;

from data routes and data centres to platforms, AI models, applications, programming languages, and productivity software—ownership, standard-setting, and strategic direction sit largely beyond India's control. Indian participation is deep, yet authority and authorship are effectively non-existent.

This concentration of control creates asymmetric leverage we have explored throughout this book. Those who own the foundational layers determine how value is created, captured, and constrained upstream. India participates intensively at the surface—through use, scale, and volume—without corresponding influence over the architectures that govern these systems. Technological dependence is therefore not occasional or circumstantial. It is as before, structural.

Within this arrangement, India's role narrows to that of being a market. Its scale generates immense value—user data, transaction volume, advertising revenue, behavioural insight—but that value flows outward.

As digital infrastructure weaves itself into daily life, every transaction, interaction, and habit generates revenue, intelligence, and learning for external owners. Platforms learn continuously about Indian consumers, businesses, and institutions. The learning is asymmetrical. India becomes legible to the West, while gaining little knowledge or capability about itself in return.

The result is a four-fold bind. Rent is extracted continuously through technologies that daily life can no longer function without. Dependence intensifies as innovation defaults to import rather than origination. Behaviour is observed, shaped, and optimised by systems designed elsewhere. And despite intense participation, no comparable strategic advantage is accumulated in return.

Yet this arrangement is routinely celebrated. Market size is framed as strategic leverage. Investment announcements are welcomed as votes of confidence. The arrival of foreign platforms, data centres, and technology firms is treated as partnership—as though capital were arriving out of goodwill rather than calculation.

But profit is not incidental to Western interest; it is its organising principle. It always has been since as early as 1600, as seen in earlier chapters of this book. Once that is acknowledged, the question becomes unavoidable: what, precisely, is India expected to receive in return?

If Western firms are eager to access the Indian market, monetise its scale, and embed themselves deeper into its infrastructure, why is this enthusiasm interpreted as collaboration rather than recognised as opportunity seeking? And if the underlying motive is extraction, what does celebration without leverage actually secure?

Part II

RECKONING WITH THE PRESENT

10

THE RECKONING

10.1 IS INDIA BEING COLONISED AGAIN?

This book has traced colonisation as a sequential process rather than a collection of isolated episodes. Each chapter examined successive stages in its evolution: from trade and dependency to monopoly; from communication control to cultural redesign; and from humiliation as a mode of subjugation to behavioural governance and consumerism. The purpose was not archival, but diagnostic: to reveal the process beneath the events, the template through which colonialism unfolds.

By identifying this structure, its flow, logic, and internal progression, the aim has been to understand colonialism not merely as a historical injury, but as a recurring pattern. To recognise what colonialism is, rather than only what it once did, is to learn from experience rather than remain captive to it. Such recognition is a prerequisite not for resentment, but for vigilance.

As seen in the previous chapters, colonialism did not operate through a single instrument. It functioned as a system. Its strength lay not in conquest alone, but in the way its components reinforced one another: how trade hardened into monopoly, monopoly into law, law into legitimacy, legitimacy into obedience, and obedience into internalised acceptance. The empire endured not because it conquered India once, but because it reorganised India, its institutions, its society, and its psyche, to reproduce its own subordination.

The question that follows is this: after the British left and India gained political independence, what exactly ended, and what remained?

Colonialism did not leave behind only railways, courts, and administrative offices. It left behind ways of seeing, choosing, and valuing. It trained instincts that favoured alignment over authorship, safety over experimentation, and adoption over creation. It expanded activity while hollowing out agency. Independence inherited institutions, markets, and infrastructure, but it also inherited the deeper patterns governing how these were used.

What remained unresolved was the capacity to author one's own destiny: the ability to imagine a future, to innovate, and to determine the direction and terms of national progress.

This capacity, authorship itself, functions like a muscle or skill. In India's case, it was not meaningfully developed in the period following independence. As a result, India entered the modern era on a compromised footing, a condition that has shaped its trajectory ever since. The modern era therefore did not begin on an empty slate. It began within inherited structures, inside a civilisation trained to participate in systems it did not design and fluent in operating frameworks it did not own and rewarded for optimisation rather than origination.

The digital age, built on semiconductors, the internet, data centres, platforms, applications, and artificial intelligence, stands as the successor to the industrial revolution that once enabled India's colonisation. These technologies do not merely represent a new phase of progress; they constitute a new set of foundational systems through which power, value, and dependency are organised.

India, shaped by its colonial inheritance and a post-independence failure to develop institutional and technological authorship, has entered this era as a participant rather than an architect, quietly mutating back into the patterns of dependency colonialism once imposed.

The preceding chapters have traced these parallels in detail. It is in this context, and only in this context, that the central question of this book must be asked carefully and without theatrics. *Is India being colonised again?*

At first glance, the question appears exaggerated. India is politically sovereign, militarily capable, economically large, and technologically active. Colonialism, after all, is commonly assumed to require soldiers, governors, flags, and explicit domination. None of these are present. To raise the question at all is therefore often dismissed as pessimism, nostalgia, or ideological excess.

This reaction, however, rests on an outdated understanding of colonialism, one that equates domination with physical occupation and mistakes visibility for control. Colonialism was never defined solely by the presence of soldiers or governors. These were instruments, not its essence. Its defining feature lay in control over foundational systems: how value was produced, who set the rules, where rents accumulated, and whose interests shaped the direction of economic and social life. Once such control is secured structurally, formal occupation becomes unnecessary.

Crucially, power of this kind rarely announces itself. The most durable forms of domination are those that are not experienced as domination at all. When control is misrecognised as efficiency, convenience, or progress, resistance appears irrational rather than necessary. As Pierre Bourdieu observed, power endures most effectively when it is misrecognised, when structures of domination present themselves as neutral systems and inequality appears as competence or merit.

It is therefore necessary to revisit colonialism not as a sequence of historical episodes, but as an evolving system. A focus on its visible instruments alone obscures the mechanisms through which domination persists after formal rule recedes. To search for colonialism in the return of kings, viceroys, or overt coercion is to misunderstand its logic.

The preceding chapters were organised accordingly: to trace the process by which trade hardened into monopoly, monopoly into law, law into legitimacy, and legitimacy into internalised compliance, culminating in enduring regimes of rent extraction and structural dependence on foreign-owned systems.

Colonialism's governing logic lay not in spectacle or administration, but in leverage through technology, control over foundational systems, extraction of value at scale, asymmetry of authorship, and the gradual internalisation of dependence. Once these conditions were secured, formal occupation became unnecessary.

Crucially, the absence of formal markers of colonial rule does not signal the absence of rule itself. The error lies in seeking colonialism in its visible expressions, in what was declared and enacted, rather than in the operating logic through which it functions. Colonialism is misrecognised by its markers rather than by its process. The disappearance of flags, administrators, and overt coercion is celebrated as independence, while process, design, and control pass largely unexamined. Authority continues to operate through structure, systems, and process, shaping economic life, regulating behaviour, and directing value flows without the need for overt command. Here, domination does not disappear; it recedes into design.

This becomes especially relevant in the digital age, where control no longer operates primarily through territory. It operates instead through infrastructure, standards, platforms, capital flows, and cognitive architectures. Under these conditions, sovereignty is not a matter of controlling land, but of controlling digital systems. Power lies less in who formally governs or occupies visible authority, than in who owns foundational layers, defines defaults, and captures learning at scale.

When the locus of control shifts in this way, the question is no longer whether colonialism can reappear, but how its logic extends into digital systems. It is this extension of structure rather than symbol that demands a new diagnostic lens.

A simple test clarifies the issue. To understand where power resides within technology-driven systems, two sets of questions must be asked: one to identify where control is concentrated, and another to identify where dependence accumulates.

Lens I: Control and Authorship

- Who owns the foundational layers?
- Who sets standards and defaults?
- Who collects rent at scale?
- Who learns continuously from usage?
- Who shapes behaviour, aspiration, and legitimacy?

Lens II: Participation and Dependence

- Who must adapt to systems they do not control?
- Who pays recurring rents without accumulating ownership?
- Who generates data, value, and learning without capturing them?
- Who bears switching costs and exit risks?
- Who experiences progress as usage rather than authorship?

Answer these questions objectively across semiconductors, cloud infrastructure, artificial intelligence models, platforms, and data flows. The outcome is clear. In India's case, participation is widespread while authorship remains scarce and largely external. Usage is massive; control is concentrated. Scale exists; sovereignty does not.

This configuration is not universal. Other states, most notably China and, through different institutional paths, the European Union, Japan, and South Korea, have made deliberate efforts to retain authorship over key technological, industrial, or regulatory layers. These efforts vary in form and success, but they share a common premise: in the digital age, sovereignty requires control over foundations, not merely participation within them.

This does not mean India is colonised in the classical sense, nor does it imply inevitability. The concern is narrower and more unsettling: India may be operating within structures that reproduce colonial patterns without being recognised as such.

A society may influence outcomes without authoring architectures. It may succeed within systems it does not control, optimise within frameworks it did not design, and grow inside structures whose direction it cannot alter. It is in this gap between operating outcomes and authoring architectures that sovereignty is quietly lost.

Historically, this imbalance was decisive. Colonial domination did not begin when foreign powers arrived with force; it began when participation expanded faster than ownership—when local actors generated value, demand, and scale within systems whose design, rules, and rents accrued elsewhere. Control followed not from exclusion, but from asymmetric inclusion.

What is striking in the present moment is how closely this pattern reappears. Once again, an asymmetry has emerged between use and authorship: participation expands rapidly, while control over foundational systems remains concentrated elsewhere. Value is generated locally, but ownership, rule-setting, and learning accumulate externally. Dependency forms not through exclusion, but through deep and continuous use.

While the asymmetry mirrors earlier periods, what is distinctive this time is how little concern it generates. The imbalance between use and authorship no longer registers as a problem. It is normalised, defended, and often celebrated. Dependency today appears rational, voluntary, and profitable. Platforms deliver efficiency. Systems reduce friction. Adoption of external technologies accelerates growth. Each individual choice appears sensible; collectively, however, these choices reproduce structural asymmetry.

What this reveals is the persistence of imperial logic, which continues to structure and shape India's technological landscape. Colonialism ended as a visible regime, but it did not fully end as a way of organising value, authority, and aspiration. The habits it produced—preference for adoption over authorship, alignment over experimentation, participation over control—were never fully unlearned.

When power later reappeared in structural form, embedded in technology rather than administration, it encountered little conceptual resistance. It was not recognised as domination because it no longer arrived as command. It arrived as infrastructure. It arrived as standards. It arrived as platforms to be joined, systems to be integrated, and efficiencies to be embraced. What had once required coercion now reproduced itself through voluntary participation.

One reason no alarm bells ring is that success is increasingly measured through externally defined benchmarks. Economic and technological performance is evaluated using indicators designed for systems that already assume ownership: market size, startup volume, platform participation, user growth, and global visibility. These adoption indicators function effectively in contexts where foundational layers are domestically owned. In such settings, adoption scale reinforces sovereignty. Other countries therefore complement adoption and growth metrics with sustained measurement of ownership, original research, intellectual property, and authorship over core systems.

In India's case, metrics are celebrated selectively. Adoption is systematically measured and rewarded, while authorship receives far less systematic attention. India undoubtedly leads globally on adoption. What remains largely unmeasured are the indicators that would reveal dependency: control over foundational systems, ownership of intellectual property, depth of original research, standard-setting authority, and accumulation of learning. What is counted reinforces participation; what is not counted conceals dependence.

At this point, the question posed by this book no longer rests on analogy or alarmism. It rests on structure. When participation expands faster than authorship, when adoption outpaces ownership, and when success is measured without regard to control, the conditions that once enabled colonial extraction begin to reappear—regardless of intent.

The central question, Is India being colonised again?, must therefore be approached not as a historical provocation, but as a functional inquiry. The issue is not whether India will be colonised as it was before, but whether the process of colonisation is unfolding again through the same underlying mechanisms.

If colonialism was once a system that used technology to convert activity into dependence, then the relevant task at this point is to ask whether a similar conversion is underway today. This question must be answered in light of an important clarification: the visible markers of historical colonialism—overt occupation, declarative rule, and coercive extraction—are no longer relevant to how power operates today.

If the answer to this question is yes, the implications for India are severe. History offers no shortage of evidence for what such conditions ultimately produce, and it would be imprudent to assume immunity from patterns already lived through. Even without revisiting that record in detail, the analysis developed across the preceding chapters leads to a clear conclusion: a civilisation financing another civilisation's future—through data, attention, consumption, and compliance—while mistaking participation for power.

The digital and artificial intelligence age intensifies this logic dramatically. Extraction now extends beyond labour and resources into behaviour, attention, preference, and decision-making itself. Rent is no longer collected episodically, but continuously. Learning systems improve with use. Surveillance is persistent rather than intermittent. Dependence deepens precisely through participation. The scale and precision of this extraction exceed anything the colonial state could have imagined. Under these conditions, the question cannot be brushed aside.

If the mechanisms are familiar, the incentives unchanged, and the asymmetry intact, only technologically amplified, on what grounds can the future be declared fundamentally different from the past?

But the greater danger lies not in answering yes. It lies in refusing to ask the question at all.

To dismiss the question is to refuse diagnosis. It denies responsibility and preserves a sense of comfort until reality asserts itself. This refusal leads to policy theatre, surface reforms, startup fetishism, and symbolic victories that leave underlying structures untouched. It creates the illusion of progress even as dependency deepens quietly beneath it.

If India is indeed drifting toward a colonial condition and refuses to recognise it, then it will not have been defeated by force or conspiracy. It will have ignored its own trajectory and surrendered control over its fate without contest.

If India fails to ask this question while time remains, it forfeits whatever chance remains to change course. That is abdication. And the cost of such abdication is not merely economic or strategic. It is civilisational and generational.

The purpose of this reckoning is not panic but recognition. Only once the condition is named can the harder work begin: identifying sources of dependency, reclaiming authorship, and designing a future that is not merely inhabited, but owned.

At the very least, recognition restores agency. It allows a society to prepare, to contest, and to play the hand it has been dealt rather than surrender by default. To acknowledge a possible future is not defeat or doom; it is the prerequisite for resistance. To look the future in the face, rather than avert one's gaze, is the first act of sovereignty.

11

THE HIERARCHY OF INNOVATION

If the analysis in the preceding chapters is correct, then the challenge before India is not merely one of growth, participation, or technological adoption. It is a challenge of authorship. The question is no longer whether India uses modern technologies, but whether it meaningfully shapes the systems through which those technologies evolve, learn, and extract value.

This distinction matters because innovation is often discussed as a single phenomenon. It is celebrated as an abstract good, invoked as a national aspiration, and measured through visible outputs such as startups, patents, or market valuations. Yet innovation is not a uniform activity. It exists in qualitatively different forms, each with very different implications for sovereignty, power, and long-term control.

To speak meaningfully about reclaiming innovation, about owning technology rather than merely deploying it, we must first be precise about what kind of innovation is being produced, where it sits in the chain of creation, and what forms of authority it confers. Without such clarity, calls for innovation risk becoming rhetorical substitutes for progress. They energise and inspire, but ultimately mislead.

The same activity that appears innovative at one level may entrench dependence at another. Some forms of innovation deepen autonomy, while others refine participation within systems designed elsewhere. Without a clear framework, success stories blur into false equivalence, heroic exceptions are mistaken for systemic health, and imitation is celebrated as sovereignty.

For this reason, before examining India's successes, failures, and prospects in innovation, it is necessary to step back and ask a more fundamental question. Not whether India innovates, but the nature of the innovation it undertakes. Below, I introduce a hierarchy of innovation that distinguishes between invention, integration, adaptation, imitation, and compilation. This hierarchy provides the conceptual lens through which India's innovation trajectory is analysed.

11.1 THE HIERARCHY OF INNOVATION

Not all innovation is equal. It can range from radical and disruptive invention to mere imitation. Informed by established theories in innovation studies, I present a hierarchy of innovation that offers an analytical basis for understanding the different levels of originality across nations and firms.

At its highest order stand the inventors — those who create products, systems, or ideas that redefine behavior, open new markets, and set the rhythm for decades to come. As Schumpeter noted, such ventures embody creative destruction — the process through which new ideas displace old orders. When successful, they shape the technological frontier and often confer the kind of monopolistic power discussed in earlier chapters.

Examples of such destructive innovations familiar to most include the semiconductor, which redefined computation and communication; the internet protocols, which became the invisible architecture through which all digital exchange now flows; the operating system, which evolved into the gatekeeper of digital life; and today's frontier technologies, such as artificial intelligence, which are beginning to extend that control from infrastructure to cognition itself.

Below destructive innovators are the integrators, who take new and successful innovations or concepts that inventors create and combine them with established technologies or ideas to deliver novel functionality or scale. Their breakthroughs lie in architecture rather than invention: they depend on prior discoveries yet create operational leaps, scalability, and ecosystem advantage. The iPod is a classic example. Apple did not invent the MP3 format, portable music players, or hard drives. What it did was integrate these existing technologies into an elegant product linked to a new distribution system, iTunes, transforming not just a device but an entire industry. Similarly, Tesla's early achievements lay not in inventing electric propulsion but in integrating battery technology, software, and manufacturing into a unified industrial architecture.

A variation of the same impulse appears in the adaptors, who localise or re-engineer global concepts to fit local economic, cultural, and infrastructural realities. They build contextual relevance and accessibility, but rarely redefine the technological frontier itself. India's Aadhaar and UPI exemplify this mode. Both took existing global ideas of digital identity and electronic payments and re-authored them for India's conditions of scale, diversity, and governance. Their innovation lies in structural redesign. Adaptation here does not occur at the surface layer of features or interfaces. It operates at the level of system architecture, reshaping the underlying logic, workflows, and constraints through which the system itself functions. UPI, in particular, was not a better app layered onto existing banking systems. It introduced a new public payment rail. In doing so, it altered the grammar of payments itself to suit the Indian landscape.

The same design instinct shaped India's Electronic Voting Machine. Voting machines existed elsewhere, but India redesigned the concept for its own constraints: offline operation, tamper resistance, multilingual interfaces, low unit cost, and deployment at civilisational scale. These were not cosmetic modifications. They were structural redesigns.

These systems demonstrate how adaptation, when applied at the level of architecture rather than interface, can still produce transformative outcomes.

Further down are the imitators, who replicate proven models to enable rapid catch-up. Much of India's early app economy belongs to this order: ride-hailing, e-commerce, and delivery platforms that mirrored Western counterparts with minimal conceptual change. They accelerated digital adoption but rarely created new intellectual or technological frontiers, serving instead refinements of global models tuned for Indian realities of price sensitivity, population density, and logistics. Their innovation lay in scale, localisation, and execution, not in original technological creation or adaptation.

Finally, there are the compilers, who assemble what is already known into efficient but familiar forms, excelling in delivery rather than discovery. India's IT services, outsourcing, and logistics networks exemplify this order: industries built on operational mastery, precision, and scale rather than original creation.

Together, these tiers define not only how nations innovate, but where they stand in the global chain of creation. Civilisational sovereignty requires sustained authorship at the upper tiers of innovation, not episodic success or heroic exception.

12

HOPE, HEROES, AND EXCEPTIONS

If the preceding chapters are correct in their assessment that India is being colonised again, then the most dangerous mistake at this point would be denial. The second most dangerous mistake would be false reassurance.

India does innovate. This is not in dispute. Across science, medicine, engineering, and enterprise, India has repeatedly demonstrated the ability to solve difficult problems, operate complex systems, and execute at scale. Any account that denies this capacity would be unserious, and any argument that rests on caricature would fail immediately.

Yet recognition alone is not analysis. The question is not whether innovation exists, but how it appears, under what conditions, and what its appearance reveals about the system that produces it.

This chapter is therefore not an audit of India's innovation ecosystem. It does not attempt to measure institutional quality, assess policy effectiveness, or rank sectors. That task would require a different instrument and a different book. The purpose here is narrower and diagnostic. It is to examine how India's genuine successes are experienced, narrated, and understood — and to ask what their exceptional character reveals about the underlying structure.

Hope, when disciplined, clarifies. When undisciplined, it misleads. In the context of innovation, hope becomes dangerous when it mistakes exception for system, and heroism for health.

12.1 INNOVATION AS EVENT, NOT EXPECTATION

In societies where innovation is systemic, success is expected. Breakthroughs occur, they are absorbed into institutional memory, and attention moves on. In societies where innovation is episodic, success must be marked, narrated, and elevated. It becomes an event.

This difference matters. When innovation functions as expectation, it does not require mythology. When it functions as exception, it does. Celebration becomes necessary not because the achievement lacks merit, but because repetition cannot be assumed. The emotional weight attached to success becomes inversely proportional to its reproducibility.

This pattern recurs across India's most frequently cited examples of innovation. What unites them is not merely achievement, but the way achievement is framed — as proof of latent greatness, as civilisational vindication, as reassurance against deeper unease. These narratives are understandable.

They are also revealing. A civilisation that celebrates every breakthrough as a miracle is quietly admitting that the system is hostile to such outcomes.

12.2 HEROES & THE CONDITIONS THAT PRODUCE THEM

The hierarchy of innovation introduced in the previous chapter offers a lens through which these successes can be examined without sentiment or dismissal. It allows us to ask not whether an outcome is impressive, but where it sits, what made it possible, and whether it can be reproduced under ordinary conditions.

ISRO and the Conditions of Exceptional Success

Consider ISRO, frequently cited as evidence of India's technological capability. Its achievements are real, sustained, and globally respected. Yet within the hierarchy of innovation outlined earlier, they largely occupy an upper integrative level rather than the technological frontier itself. ISRO has demonstrated the ability to assemble, coordinate, and execute complex systems with discipline and efficiency, but it has rarely been tasked with — or permitted to pursue — frontier-defining innovation.

A defining feature of ISRO's work is its cost profile. Its missions are consistently executed cheaply relative to those of frontier first movers. This is often treated as a virtue

— and in operational terms, it is. But analytically, it is revealing. Frontier innovation is almost never cheap. First movers operate in conditions of radical uncertainty, prolonged experimentation, repeated failure, and sustained capital burn long before success becomes visible. High cost is not a flaw of frontier innovation; it is one of its structural signatures.

ISRO's ability to deliver credible outcomes within tightly constrained budgets therefore signals its position in the hierarchy. What it demonstrates is not frontier authorship, but excellence in integration, optimisation, and disciplined execution within known technological trajectories. The very affordability that attracts admiration also explains the nature of the achievement.

This, in turn, explains the celebration that surrounds ISRO. Its successes are treated as civilisational events, proof points of latent greatness, moments of national reassurance. That reaction is understandable — but it is also diagnostic. In systems where frontier innovation is routine, success is expected and quickly absorbed, attracting little ceremony. Where innovation appears episodically and below the frontier, it must be elevated, narrated, and mythologised. Celebration compensates for the lack of repeatable, expectational frontier innovation.

At the same time, ISRO's record reveals a clear ceiling. Its mandate appears to have been shaped by strategic restraint and diplomatic caution, consistent with a broader national emphasis on demonstrating capability and credibility rather than redefining technological frontiers. This helps explain why its missions remained aligned with trajectories already established elsewhere. Such alignment was not incidental; it functioned as a condition of acceptance and legitimacy. ISRO was not asked to invent new propulsion paradigms, redefine materials science, or open fundamentally new technological frontiers. It was asked to operate competently, credibly, and economically within existing ones. Integration, however sophisticated, does not confer the same strategic leverage as invention. It enables participation without authorship, presence without agenda-setting.

And yet, within those limits, ISRO does offer genuine hope — precisely because its success was not accidental. ISRO was deliberately granted leeway that is largely absent across India's broader innovation ecosystem. It was insulated from short-term market logic, shielded from platform rent extraction, and afforded long time horizons under a sovereign mandate. Most importantly, it operated under a logic of belief before proof rather than proof before belief. It was allowed to fail early, to learn without immediate justification, and to continue receiving support before success was demonstrated rather than after it was guaranteed.

That leeway mattered. ISRO shows what becomes possible when structural constraints are relaxed and institutions are trusted to mature before they are judged. But it also clarifies the broader diagnosis: even India's most protected technological institution has largely been positioned to integrate, optimise, and execute rather than to originate. That is not a failure of talent. It is evidence of a system that does not yet author its own frontiers.

Jio and Scale as a Proxy for Innovation

The same pattern appears, in a different form, in the private sector. Reliance Jio is often described as a triumph of Indian entrepreneurship, and in important respects it is. Jio transformed internet access in India, collapsed data prices, and accelerated digital participation at civilisational scale. It solved a distribution problem with extraordinary resolve and speed.

What is striking, however, is how this transformation occurred. India did not become digitally accessible through a systematic, institution-led expansion of foundational infrastructure. It became digital because a single private actor chose to assume the burden. The trigger was contingent, not structural — famously articulated when the daughter of the actor remarked on India's digital experience relative to the West. That observation mattered not because it was analytically unique, but because it was voiced by someone with the capital, authority, and risk tolerance to act on it.

In societies where digital access is treated as foundational, the development of digital infrastructure cannot depend on individual conviction or exceptional will. It is planned, financed, and delivered as a matter of course. In India's case, digital inclusion advanced because one entrepreneurial decision substituted for institutional absence. What later came to be celebrated as Digital India was not the outcome of an intentionally designed public architecture, but the result of private initiative, capital concentration, and individual resolve stepping into an institutional vacuum. Jio was not the product of a system working as designed; it was a workaround for one that had not been designed at all. What is worse is that it is celebrated for being non-systematic.

Within the hierarchy of innovation, Jio's achievement sits firmly at the level of integration and execution, not technological authorship. The foundational layers—hardware, network technologies, and core software architectures—remained largely imported. Control resided upstream. Value capture accumulated elsewhere. What Jio mastered was scale, pricing, logistics, and rollout within an external technological stack. Distribution was mistaken by many for sovereignty.

And yet, the achievement should not be minimised. Jio's intervention unlocked a voracious appetite for digital services across India. It enabled platforms, payments, content, commerce, and connectivity to proliferate at unprecedented speed. Entire sectors expanded on the back of that access. In this sense, Jio revealed something real: latent demand was never the problem. Capability to consume, adapt, and build atop known systems existed in abundance.

This is where the hope lies, and where the warning must be heeded. Jio shows what becomes possible when access constraints are removed. But it also exposes the fragility of relying on episodic, heroic intervention to deliver foundational change. A society cannot depend on chance conviction, family fortune, or individual resolve to build its infrastructure. What was achieved through exception must become routine. What arrived through heroism must become expected.

Until such outcomes are systematised — institutionally planned, structurally financed, and repeatably delivered — they will continue to appear as miracles rather than norms. And a civilisation that waits for miracles to build its foundations is not short of talent or ambition. It is short of architecture.

Covaxin and Innovation Under Emergency Conditions

Crisis-driven successes follow a similar logic, though in a more compressed and revealing form. Covaxin emerged under extraordinary pressure, with accelerated regulation, direct state backing, and emergency mobilisation. Its contribution is undeniable. Covaxin delivered when it mattered, helped stabilise India's vaccination response, and contributed meaningfully to vaccine availability across the Global South, at a moment when access was sharply unequal.

What made this possible was not normal institutional functioning, but the temporary suspension of it. COVID-19 disabled many of the bureaucratic stops, procedural delays, and risk-avoidance norms that ordinarily govern high-end biomedical research in India. Research priority was elevated, funding and regulatory pathways were compressed, and state support was extended before outcomes were fully known.

Even so, pressures to import, defer, or wait for external validation reportedly persisted, underscoring how exceptional the operating conditions remained, even during national emergency.

Within the hierarchy of innovation, Covaxin represents high-end applied research executed under crisis insulation, rather than a routinely reproducible research pipeline. The scientific competence involved was real. The mobilisation was effective. But the success was conditional on emergency authorisation, political urgency, and global crisis, not on a standing ecosystem that reliably produces frontier-adjacent biomedical innovation under normal conditions.

The ceiling becomes visible when the context is examined more closely. While execution and development occurred domestically, the problem itself was externally defined. COVID-19 was not a research question posed by Indian scientific institutions in advance, nor the outcome of long-term sovereign agenda-setting in biomedical science. It was a global emergency, framed, prioritised, and validated within a Western-dominated research and regulatory landscape. Indian science responded decisively, but it did not set the question.

That distinction matters. Responding effectively to externally defined problems demonstrates competence and capacity. Defining problems before they become crises, and sustaining research where no immediate validation exists, is what characterises scientific authorship. Covaxin shows that India can deliver under pressure. It does not yet show that India systematically pursues high-risk, long-horizon research where the question itself originates domestically.

And yet, the achievement should not be minimised. Covaxin demonstrates that when insulation, priority, and belief precede proof, Indian institutions can operate at a high level of scientific execution. The hope lies precisely here, not in the crisis itself, but in what the crisis temporarily enabled. The warning is equally clear. An innovation system that requires emergency to function is not resilient. What was achieved under exception must become normal. What emerged in response must be made anticipatory. Until that shift occurs, India will continue to innovate when forced, rather than because it has designed itself to do so.

12.3 THE STARTUP ARGUMENT, REVISITED

Much of India's remaining innovation activity is channelled through startups and service providers. This energy is real and economically valuable, but it is shaped by the incentives under which it operates. Predominantly governed by venture capital and market logics, startup innovation is pulled toward problems that are fast to validate, quick to scale, and commercially legible. Service providers operate under a similar discipline, optimising for reliability, cost, and client demand within problem definitions set elsewhere.

This incentive structure reinforces a familiar constraint: proof before belief. Ideas are funded, staffed, and scaled only once demand, credentials, or traction are already visible. As a result, startups and service providers tend to occupy lower levels of the innovation hierarchy, not because of limited competence, but because they are structurally oriented toward low-risk, low-uncertainty domains. Integration, localisation, and execution dominate; frontier uncertainty is systematically avoided.

Such a model produces capable execution, employment, and incremental growth, but it also narrows ambition. Frontier technologies that are foundational, slow to mature, capital-intensive, or strategically necessary but commercially uncertain are filtered out early. Startups and service providers excel at responding to opportunity; they cannot be expected to define national technological direction. Left to market randomness, innovation clusters where returns are easiest, not where capability is most needed.

The implication is not that startups or services are inadequate, but that the prevailing innovation model itself requires an upgrade. Entrepreneurial energy must be complemented by state leadership that defines long-horizon technological priorities, supplies patient capital, and absorbs early risk. Without this, even successful ventures remain dependent on external infrastructure and foreign foundations. With it, startups and service providers can shift from opportunistic participation to structured contribution. National technological capability cannot be built by accident or heroics; it must be planned for with foresight.

Finally, what remains are isolated scientific breakthroughs: papers, prototypes, or laboratory results. These do not contradict the diagnosis advanced here; they exemplify it. Discovery without institutional continuity is a symptom of episodic capability, not systemic innovation. Once such breakthroughs occur, they are often amplified, celebrated, and elevated as proof of national potential, not because they are unprecedented, but because they are rare.

In systems where innovation is systemic, such events attract far less attention. Discoveries are expected, embedded within tightly coupled research networks, programmatic funding structures, and institutional pathways that carry results forward into further inquiry, application, and replication. There, breakthroughs are not treated as validation of possibility; they are routine outputs of an ecosystem designed to produce them.

The contrast is revealing. What is celebrated as exceptional in one context is unremarkable in another, not because of differences in talent, but because of differences in expectation and structure. Where continuity exists, novelty does not require narration. Where continuity is absent, novelty must be mythologised.

12.4 MYTH AS SYMPTOM

What ultimately distinguishes episodic innovation from systemic innovation is not talent or intent, but expectation. Where systems function, innovation is normalised. Where systems are absent or misaligned, sporadic innovation must be elevated, narrated, and mythologised.

This myth-making is not deception. It is compensation. It fills the gap between capability and structure. It reassures where the repetition of breakthrough innovation cannot yet be guaranteed.

The danger lies not so much in pride but in misrecognition. When heroes are taken as evidence that the system works, the harder work of system-building is deferred. Celebration substitutes for design. Hope replaces architecture. Hope itself is not the problem. Hope that relies on chance, as though progress might fall from the stars, is.

12.5 REFRAMING HOPE

Properly understood, this chapter is not pessimistic. It is clarifying. It does not deny India's capacity to innovate; it explains why innovation so often appears as exception rather than routine.

Hope does not lie in producing more heroes. Hope lies in redesigning the conditions that make heroes unnecessary.

If innovation continues to require insulation, emergency, or extraordinary resolve, the problem is not talent but structure. Innovation is never the product of a single factor. It emerges only when multiple elements align: national priorities that define what must be built, institutions that reward long-term risk, capital that tolerates uncertainty, infrastructure that carries ideas forward, and a societal mindset that treats experimentation as normal rather than reckless.

What is needed, therefore, is not encouragement but architecture, a systemic approach that begins by asking what the country actually needs, rather than what happens to be profitable, fashionable, or immediately defensible. An ecosystem in which innovation is not hoped for, celebrated, or mythologised after the fact, but expected by design. Such an ecosystem cannot be assembled piecemeal. It requires institutional coordination, cultural reorientation, and deliberate engineering across multiple layers of society.

That task lies beyond critique alone. It requires a framework.

Before turning to that framework, however, it is instructive to examine how such alignment has succeeded in one civilisation and failed in another. The next chapter does so through an earlier material foundation of power: iron. Once mastered by India and later lost, iron offers a historical lens through which the mechanics of innovation, ecosystem collapse, and systemic coherence can be understood, and through which the deeper requirements of renewal come into view.

Part III

LOOKING TO THE FUTURE

13

IRON: THE GAIN & LOSS OF THE INNOVATION SPIRIT

> *Gold is for the mistress, silver for the maid,*
> *Copper for the craftsman cunning in his trade.*
> *"Good," said the baron sitting in his hall,*
> *"But iron—cold iron—is master of them all."*
>
> — Rudyard Kipling

This verse by Rudyard Kipling is recited by Canadian engineers as part of the Iron Ring Ceremony, an oath to uphold ethics and restraint in the use of power. It reminds engineers that mastery must remain humble, that creation and destruction often share the same forge. The story of iron is, on reflection, the story of power itself.

13.1 INDIA AND THE AGE OF STEEL

Since the third century BCE, India was the master of steel. Its furnaces made wootz—high-carbon, consistent, and pure. Hyderabad, Mysore, Salem, and Central India were to metallurgy what Silicon Valley is to code. Each region refined its own method, but the results were the same: blades that could cut iron, springs that never lost tension, and surfaces etched with artful patterns. Arab traders carried the ingots west across the Indian Ocean, where European smiths forged them into the legendary Damascus blades. For nearly two thousand years, no one could match their quality.

Even Michael Faraday, the father of electromagnetism, tried and failed. Between 1818 and 1822, working at the Royal Institution in London, he conducted hundreds of experiments to reproduce Indian crucible steel. He could not replicate its purity, strength, or pattern.

Yet the spirit of innovation is never permanent; it cannot be assumed, inherited, or taken for granted. A chain of economic, political, and geographic events soon dislodged India's two-thousand-year dominance of steel. Within decades, its furnaces went silent. The knowledge that had passed from hand to hand since the Mauryan age vanished. In its place rose British steel—cheaper, faster, and protected by law—the metal that powered the Industrial Revolution.

Innovation never manifests by a single cause. It happens only when multiple systems, or layers, align: institutions that reward risk, networks that scale production, and a societal mindset that values experimentation. When any of this fractures, innovation potential collapses. The story of iron offers a glimpse of how these layers converged in Britain and remained undeveloped and unsynchronised in India, giving rise to the Industrial Revolution in one and silence in the other.

13.2 THE ENGLISH PROBLEM

By the early eighteenth century, England faced a problem buried underground. Its coal mines were flooding. The deeper they went, the more water they met. Horses and manual pumps could not keep up, shutting down mines and threatening the nation's main source of fuel.

In 1712, a Devonshire ironmonger named Thomas Newcomen built a machine to solve it. His steam engine used heat and pressure to drive a piston that pumped water from the pits. It was large, slow, and inefficient—but it worked. The engine could be copied, rented, and repaired. There was voracious demand for it across the coalfields.

Innovation thrives when it begins in necessity, not imitation. The closer an invention sits to a real problem, the greater its chance of becoming valuable to society.

13.3 THE ECOSYSTEM THAT MADE IT POSSIBLE

The invention of the steam engine alone did not create industrialisation. When backed by capital, law, and institutions—and by a society that rewarded invention—something larger began to move. The Statute of Monopolies (1624) gave inventors the right to profit from their ideas. Joint-stock companies, private banks, and merchant investors, drawn by the extraordinary returns of mechanisation, pooled capital to fund efficiency.

What emerged was an institutional environment that rewarded initiative, protected profit, and financed progress. When risk is rewarded rather than punished, innovation compounds. Soon, an engine built to drain coal mines found use far beyond them. It powered textile mills, ships, and factories—and improved with every application and iteration.

Each factory became a centre of invention—demanding customised tools and producing goods faster and more precisely than ever. Machines, mills, and foundries transformed power into production. Each factory made parts for others, forming a web of interdependence that produced increasingly complex goods at scale. Every factory was a node in this network—a fortress of know-how. Without mastery of each node, the dream of replicating—or even imitating—complex products was mere wishful thinking.

Around them grew roads, canals, ports, the postal system, telegraph lines, and railways that connected production to markets and knowledge to movement. Britain built not just machines but motion. Every new route reduced friction; every new port multiplied trade. Every new canal cut the cost of moving goods across land, turning geography itself into an advantage. Such infrastructure became the nervous system of progress, enabling both innovation and its impact.

Among these factories, one stood apart: the iron foundry. What began as a support industry for pumps and pistons became the backbone of empire. Innovation rarely follows a straight line—it often begins at the margins, where practical necessity forces experimentation and solution-finding. This spark, in turn, once it catches on, feeds its own fire.

The demand for stronger, more reliable parts for steam engines pushed Britain's ironmakers to refine their methods. The very steam engine built to mine coal became the enabler of better steel. It was used to cast larger cylinders, perfect smelting temperatures, and standardise alloys. Each improvement in machinery required better metal, and better metal enabled more powerful machines. Iron became rails; rails carried coal; coal powered engines; engines built ships. A cycle of innovation and progress had begun—each link strengthening the next.

From these furnaces came the material that destroyed India's wootz tradition—not through superior craft, but through the economics of scale. Power, as history repeatedly shows, is never in the tool itself but in the ability to reproduce it at scale—and in the suppression of competition. Once the British gained an edge, they did not rest; they pressed it. Indian furnaces were taxed, forest access for smelting restricted, and local methods branded primitive. Imports of British iron flooded Indian markets. Forest access, vital for Indian smelting, was curtailed under the pretext of protecting timber for ships and railways—a calculated move to restrict critical inputs and raise rivals' costs. These moves were deliberate and rational. In game-theoretic terms—solidifying dominance—this was strategic exploitation: when a player gains a technological lead, the logical step is to deny rivals the means to compete.

This raises an enduring question: if rational behaviour leads to dominance, is morality merely the privilege of the powerful, the refuge of the powerless, or neither?

13.4 THE SILENCE OF THE INDIAN FURNACE

By the mid-nineteenth century, the old wootz furnaces had disappeared, and the British method of steel had become dominant and remains so. The artisans who once produced the world's finest metal were outpriced, outlawed, and forgotten.

The knowledge they guarded so tightly was lost to time. However, the blame cannot rest solely on the British. India itself was primed for failure in the innovation race.

Under the Mughal, Mysore, and Maratha rulers, Indian foundries had a narrow purpose: to supply swords, armour, gates, and cannon for royal courts and armies. Orders were scattered—one sword for a prince, one door for a palace, one cannon for a fortress. Production was artisanal, not industrial. Each commission was unique and created no scale.

Indian rulers, acting as both buyers and benefactors, lacked the foresight to secure the survival of the iron industry. India's ruling and mercantile classes failed to guard its interests, blind to the strategic value of their own iron. When India's empires collapsed, the furnaces went silent.

Under British rule, India continued to supply raw ore and consume iron for railways, bridges, and ships, but the conversion of iron into machinery and infrastructure was relocated abroad. Extraction remained, consumption increased, but value creation declined.

Indian society shared the same blindness as Indian rulers. Centuries of metallurgical mastery had made it complacent. Iron was treated as a symbol of status, not a foundation of industry. The culture revered craftsmanship but never scaled it into systems. It lacked the imagination to see that the true power of iron lay not in ornament or warfare but in machinery. This failure of imagination marks the difference between societies that use innovation voraciously and those that merely inherit and preserve it.

Financing, too, was absent. British inventors had access to banks and investors who competed to fund invention. In India, craftsmen relied on royal advances or personal loans—often at arbitrary rates and subject to the whims of the powerful. When the courts fell, credit vanished. Without capital, skill alone could not survive. Capital is not merely money—it is continuity, competition, and confidence.

Knowledge remained fragmented. In Britain, discoveries circulated through patents, journals, and universities. Law protected innovators, giving them the right to profit from discovery. Entrepreneurs could experiment and fail without ruin. In India, knowledge stayed confined within families and guilds, guarded as secret inheritance. Innovation depended not on law or market, but on lineage, patronage, and rank. Secrecy preserves mastery but suffocates progress.

In contrast to Britain's thriving ecosystem of engineers and factories, India's conditions for innovation were inhospitable. A civilisation that had mastered steel no longer valued it. Two thousand years of unchallenged expertise had made iron ordinary. Its strategic importance was forgotten. The very success of India's metallurgical past denied any urgency for renewal.

Prosperity often dulls the very instinct that built it—the will to struggle, adapt, and create anew.

In Indian tradition, the goddess of prosperity is said to bless those who use wealth and resources with purpose and respect. The spark of innovation is no different: creativity is not lost—it leaves. With no respect for what iron could build, and no curiosity for what more it could do, the spirit of innovation left India after two millennia and made its home in England, where it found the recognition and ambition it deserved.

Iron, like talent, thrives where it is used.

14

TOWARDS DIGITAL SWARAJ

The story of iron shows that sovereignty is never lost in a moment. It erodes when systems fail to align, when capability is treated as heritage rather than something that must be continuously rebuilt.

The mastery of iron did not rise in Britain through a single invention, nor did it vanish from India for any single reason. In both cases, what mattered was not the event, but the coherence of the systems that surrounded it. Britain aligned institutions, capital, infrastructure, knowledge circulation, and social incentives around innovation, production, and scale. India, despite two millennia of metallurgical mastery, did not. When circumstances shifted, Britain compounded advantage. India absorbed the shock and consequences.

This pattern is not unique to iron. Nor is it confined to the nineteenth century. The shift from iron to computation may appear vast. Steel was heavy, visible, and territorial; digital systems are abstract, portable, and global. Yet the underlying logic of innovation in both realms remains unchanged.

Further, as seen in this book, power accrues to those who design and control the foundational systems upon which others depend. Societies that fail to author these layers do not merely fall behind; they become structurally dependent. In an interconnected world, there is no such thing as innovation in isolation. When a society innovates, its progress shapes other nations regardless of intent and irrespective of consent. Competition, therefore, is not optional. It is unavoidable.

If India is indeed drifting toward a new colonial condition, the question is no longer rhetorical. It is practical. How does a society reclaim sovereignty when dependence is embedded in infrastructure, platforms, and standards? What would it take to move from adoption to authorship? And what, then, is required to achieve Digital Swaraj?

This book has not attempted to answer those questions by design. Its task was diagnosis: to make visible the mechanisms through which dependency forms, normalises, and sustains itself. History does not instruct. It warns. How a society responds to that warning is a separate question, deserving its own diagnosis.

The book that follows is dedicated to that work.

One revelation, however, begins to reveal itself. If innovation emerges when multiple layers of society align, and power accrues to those who innovate, then power is simply a function of ecosystem maturity.

Across eras, innovative societies appear to follow a recognisable logic. That logic makes innovation not accidental, but inevitable. It also drives such societies to export their technologies, scale them globally, and embed them elsewhere spreading their own rhythms and dependencies across others. In doing so, they do not merely expand markets. They extend control, compete, and refine themselves, evolving alongside the systems they set in motion.

This tendency has a clear parallel in ancient Indian tradition, most notably in the ashvamedha, where an emperor released a horse to roam freely, claiming sovereignty over every territory it entered while others were left unable to resist. Modern technologies move in much the same way—circulating freely and asserting dominance.

Such a society does not pursue innovation merely to defend itself, to survive, or to endure. It lives for the spirit of innovation itself—as a reason to thrive, to improve, and to explore.

It is in that spirit that this book ends. Not with answers, but with a boundary. Not with reassurance, but with recognition.

The task ahead is creative and forward-looking: to define how innovation is harnessed, how it is channelled into enduring capability, and how power is ultimately organised through it.

It is to recognise the reciprocal relationship between society and technology: that a society capable of creating powerful technologies must itself be shaped in particular ways, and that those technologies, once created, go on to reshape society in return. At this stage, conscious reflection becomes essential. How we choose to inhabit the world, how we shape it, and what intentions guide that shaping can no longer remain implicit. They must be made explicit.

The aim is not merely to state or hope for the future, but to build systems that entrench a sovereign Indian future so deeply that the future itself becomes attentive to our intentions.

If India has the power to shape itself and the world, how will it choose to do so?

Towards India's Digital Swaraj

REFERENCES AND FURTHER READING

EMPIRE, TRADE, AND HISTORICAL POWER

- Landes, David S. *The Wealth and Poverty of Nations: Why Some Are So Rich and Some So Poor.*
- Mokyr, Joel. *The Lever of Riches: Technological Creativity and Economic Progress.*
- Dalrymple, William. *The Anarchy: The East India Company, Corporate Violence, and the Pillage of an Empire.*
- Guldi, Jo. *Roads to Power: Britain Invents the Infrastructure State.*

MARKETS, NETWORK EFFECTS, AND ECONOMIC LOCK-IN

- Shapiro, Carl, and Hal R. Varian. *Information Rules: A Strategic Guide to the Network Economy.*
- Varian, Hal R. "Market Structure in the New Economy."

LAW, CORPORATIONS, AND DELEGATED SOVEREIGNTY

- Chang, Ha-Joon. *Kicking Away the Ladder: Development Strategy in Historical Perspective.*
- Winner, Langdon. "Do Artifacts Have Politics?" *Daedalus.*

TECHNOLOGY, INFRASTRUCTURE, AND POWER

- Zuboff, Shoshana. *The Age of Surveillance Capitalism: The Fight for a Human Future at the New Frontier of Power.*
- Castells, Manuel. *The Rise of the Network Society.*
- Srnicek, Nick. *Platform Capitalism.*
- McLuhan, Marshall. *Understanding Media: The Extensions of Man.*
- Deleuze, Gilles. "Postscript on the Societies of Control."
- Schumpeter, Joseph A. *Capitalism, Socialism and Democracy.*
- Morozov, Evgeny. *To Save Everything, Click Here: The Folly of Technological Solutionism.*

BEHAVIOUR, MEASUREMENT, AND SOCIAL ENGINEERING

- Hobbes, Thomas. *Leviathan.*
- Quetelet, Adolphe. *A Treatise on Man and the Development of His Faculties.*

- Espeland, Wendy Nelson, and Michael Sauder. "Rankings and Reactivity."

INDIA, DEPENDENCY, AND DIGITAL INFRASTRUCTURE

- Government of India, Ministry of Electronics and Information Technology (MeitY). Selected policy documents.
- India Electronics and Semiconductor Association (IESA). Industry reports on semiconductor demand and supply-chain dependencies.
- NITI Aayog. Reports on digital public infrastructure and platform governance.
- Observer Research Foundation (ORF). *The Evolving Semiconductor Supply Chain Landscape* (2025).

COMMUNICATION, LANGUAGE, AND NARRATIVE CONTROL

- Anderson, Benedict. *Imagined Communities.*
- Naoroji, Dadabhai. *Poverty and Un-British Rule in India.*
- Gandhi, M. K. *Hind Swaraj.*

HUMILIATION, CULTURE, AND COLONIAL PSYCHOLOGY

- Fanon, Frantz. *Black Skin, White Masks.*

- Nandy, Ashis. *The Intimate Enemy: Loss and Recovery of Self under Colonialism.*
- Tharoor, Shashi. *An Era of Darkness.*

TASTE, STATUS, AND IMITATION

- Veblen, Thorstein. *The Theory of the Leisure Class.*
- Bourdieu, Pierre. *Distinction: A Social Critique of the Judgement of Taste.*

CONTEMPORARY REPORTING AND CASE MATERIAL

- Indian business media coverage of the Microsoft–Nayara digital access incident (2025).
- International reporting on platform governance, cloud dependence, and data sovereignty initiatives (Reuters, *Financial Times, Business Standard, Economic Times*).
- OECD. Reports on data governance, cloud dependence, and digital sovereignty.